SpringerBriefs on Pioneers in Science and Practice

Volume 19

Series editor

Dr. Hans Günter Brauch, Mosbach, Germany

For further volumes:
http://www.springer.com/series/10970
http://www.afes-press-books.de/html/SpringerBriefs_PSP.htm

Hans Günter Brauch · Teri Grimwood
Editors

Jonathan Dean

Pioneer in Détente in Europe, Global Cooperative Security, Arms Control and Disarmament

Springer

AFES PRESS

Editors
Dr. Hans Günter Brauch
AFES-PRESS
Mosbach, Baden-Württemberg
Germany

Teri Grimwood
Nuclear Weapons and Global Security
Union of Concerned Scientists
Washington, DC
USA

The cover photograph was taken by Mr. Steve Schiro, Amb. Jonathan Dean's son-in-law who granted permission to use it here. Tds 3.23.14

Copyediting: PD Dr. Hans Günter Brauch, AFES-PRESS e.V., Mosbach, Germany

ISSN 2194-3125 ISSN 2194-3133 (electronic)
ISBN 978-3-319-06661-5 ISBN 978-3-319-06662-2 (eBook)
DOI 10.1007/978-3-319-06662-2
Springer Cham Heidelberg New York Dordrecht London

Library of Congress Control Number: 2014937723

Printed on acid-free paper

Springer is part of Springer Science+Business Media (www.springer.com)

Foreword

Jonathan Dean was born on 15 June 1924 in New York City and passed away on 14 January 2014. On the 90th anniversary of his birth, this book commemorates the lifelong achievement of Ambassador Dean both as a distinguished American diplomat and as a global citizen, recognizing his contributions to détente policy in Europe, global cooperative security, arms control and disarmament negotiations, and public debates between policymakers, diplomats, scientists and concerned citizens in the United States and Europe.

At the age of 17, Jonathan Dean had joined the Canadian forces to help liberate Europe from the Nazi regime, and he served as an American infantry officer in the campaign from the Normandy landings to the Elbe. After World War II, he studied at Harvard and Columbia Universities and received a Ph.D. in Political Science from George Washington University. He began his Foreign Service career in 1950 in Bonn as liaison officer between the US High Commission and the Federal German government; it was there that he helped to establish the new Federal German armed forces and facilitated Germany's joining NATO.

Later he served as desk officer for East Germany in the Department of State (1956–1960) and as first secretary at the American Embassy in Prague (1961–1962). In the early 1960s, he was the principal officer in Elisabethville, Katanga (1962–1964) during the Tshombe secession and the subsequent UN peacekeeping intervention, and was also deputy director of the Office of United Nations Political Affairs, Department of State, where he worked on peacekeeping and economic sanctions. In 1968, he returned to the American embassy in Bonn as Political Counselor and deputy US negotiator for the 1971 Quadripartite Agreement on Berlin.[1]

[1] See the detailed transcript of an interview with Jonathan Dean by Charles Stuart Kennedy carried out on 8 July 1997 as part of the Foreign Affairs Oral History Project on behalf of the Association for Diplomatic Studies and Training; at: <http://www.adst.org/OH%20TOCs/Dean,%20Jonathan.toc.pdf>. This interview offers insightful comments on complex debates within the Nixon administration between the State Department and Henry Kissinger as director of the National Security Council (1969–1973), the negotiations between the four allied powers, and the controversial political debate in Germany between the government of Chancellor Brandt representing the Social and the Free Democrats and the opponents of German Ostpolitik represented by the Christian Democrats (CDU) and the Christian Social Union (CSU). See below Chap. 1 by Teri Grimwood; See also the website on this book, at: <http://www.afes-press-books.de/html/SpringerBriefs_PSP_Dean.htm>.

This Agreement was signed on 3 September 1971. It complemented several bilateral agreements between the Federal Republic of Germany and the Soviet Union ("Moscow Treaty," signed 12 August 1970); Poland ("Warsaw Treaty," signed 7 December 1970); and the "Transit Agreement" between West and East Germany of 12 December 1971. The Quadripartite Agreement on Berlin eased the way both for the multilateralization of détente in Europe resulting in the *Conference on Security and Cooperation in Europe* (CSCE), whose Helsinki Final Act was signed at the Helsinki summit of 29 July to 1 August 1975, and for the launching of the *Mutual and Balanced Force Reduction* (MBFR) talks in Vienna, where preliminary talks began in January 1973 though it was in late 1973 that the negotiations took off.

The MBFR Talks ended on 2 February 1989 and were replaced by a mandate for negotiations on *Confidence- and Security-Building Measures* (CSBM) on 15 January 1989 and by the opening of the negotiations on *Conventional Forces in Europe* (CFE) between the 23 members of NATO and the WTO. After the fall of the Berlin Wall and the ending of the Cold War on 17 November 1990, these two parallel sets of negotiations resulted in the signing of the *Vienna Document 1990 on CSBMs* at the summit of CSCE members in Paris.

Two days later, on 19 November 1990, the Treaty on *Conventional Armed Forces in Europe* (CFE) was signed by 22 NATO and Warsaw Pact countries; it was ratified on 25 November 1991 by the US Senate and on 8 July 1992 by the Russian Supreme Soviet.

On 9–10 July 1992, the CSCE Helsinki Summit approved the Helsinki Document 1992 and agreed to establish a new CSCE *Forum for Security Cooperation* (FSC), a merger of the CSBM and CFE talks, to include all 52 members, and to strengthen the Conflict Prevention Center. At this meeting, the Concluding Act of the Negotiations on Personnel Strength of Conventional Armed Forces in Europe (CFE 1A) was also signed. A week later, on 17 July 1992, the CFE Treaty entered into force.[2]

As a US negotiator for the Berlin Agreement of 1971 and as the Deputy US Representative (from 1973) and as a US Representative with the rank of Ambassador (1978–1981) to the force reduction negotiations in Vienna (the MBFR talks), Jonathan Dean played a crucial role both in the framing and conduct of the negotiations leading to the Berlin Agreement that entered into force in September 1971, and in the launching of the MBFR talks. He was thus one of the architects of a diplomatic process that contributed to a multilateralization of détente in Europe and eventually to the ending of the Cold War with its symbolic date of the fall of the Berlin Wall on 9 November 1989, some 25 years ago.

After having worked for three decades in the US Foreign Service (1950–1980), he left the State Department when Ronald Reagan became President, and in 1981 joined the Carnegie Endowment for International Peace as resident associate for arms control and European security issues; in addition, from 1984 to 2007 he was

[2] This chronology relies on a fact sheet of the Federation of American Scientists (FAS) on the "CFE Chronology: Conventional Forces in Europe Treaty", at: http://www.fas.org/nuke/control/cfe/chron.htm.

global security adviser to the *Union of Concerned Scientists* (UCS), which became his new institutional home. The UCS supported his innovative, policy-relevant, conceptual, and scientific work on issues related to détente in Europe, global cooperative security, and arms control and disarmament during the second cold war and in the post-cold war era, until he retired in 2007 at the age of 83.

Ambassador Jonathan Dean was a distinguished American diplomat and a concerned US citizen who after his retirement was associated—as well as with the UCS—with many other organizations in the United States, e.g., as President of the Washington, D.C. chapter of the *United Nations Association*; President, *Institute for Defense and Disarmament Studies* (IDDS); Research Advisory Board, *Project on Defense Alternatives*; Advisory Board, *Global Security Institute;* Director, *NGO Committee on Disarmament, Peace and Security*; Steering Committee Member, *Global Action to Prevent War*; National Advisory Board, *Council for a Livable World*, and in Europe as a Council Member, *British American Security Information Council* (BASIC), working for détente, arms control and disarmament. From 1991, he was chairman and from 2007 co-chair of the scientific advisory council of *Peace Research and European Security Studies* (AFES-PRESS), an international nonprofit scientific society based in Germany.

He authored several books: *Watershed in Europe* (1987), *Meeting Gorbachev's Challenge* (1989), and *Ending Europe's Wars* (1994), and co-authored *The Nuclear Turning Point* (1999). Dean's views were published in the *New York Times, Los Angeles Times, Boston Globe, Christian Science Monitor, Washington Post* and *Baltimore Sun,* as well as in journals such as *Foreign Policy, International Security*, and *Foreign Affairs*.

As a diplomat Jonathan Dean was actively involved in integrating Germany into the West and gradually overcoming the Cold War as the deputy US negotiator of the Berlin Agreement (1971) and as the deputy and US representative at the MBFR talks. After his retirement, he contributed conceptually to the development of policies that aimed at détente in Europe and arms control and global disarmament before and after the global turn. He was a great, engaged spirit who significantly helped to change the face of Europe and the world.

I met Ambassador Dean for the first time during the 26th annual conference of the *International Studies Association* (ISA) in March 1985 in Washington, D.C., where we had a brief conversation after one of the arms control panels. After this initial conversation a close cooperation based on mutual respect, trust, and friendship emerged. His first contribution to my German book on Confidence-Building Measures and the European Disarmament Conference (1986) was a chapter on "A Crisis Management Center in Europe: a concrete proposal for a confidence- and security-building measure for Vienna and Stockholm".[3]

[3] This chapter was published as: "Ein Krisenmanagementzentrum in Europa—ein konkreter Vorschlag einer Vertrauens- und Sicherheitsbildenden Maßnahme für Wien und Stockholm", in: Hans Günter Brauch (Ed.): *Vertrauensbildende Maßnahmen and Europüische Abrüstungs konferenz—Analysen, Dokumente and Vorschüge*, Militärpolitik and Rüstungsbegrenzung, Vol. 9 (Gerlingen: Bleicher-Verlag, 1986): 329–338.

This scientific cooperation between a distinguished diplomat and a young critic of the armament policies of the Reagan administration lasted for nearly three decades until his death.

On 28 November 2005, he sent me one of his last publications on: "Rethinking Security in the New Century: Return to the Grotean Pattern," which he wrote as a preface to a volume on *Globalization and Environmental Challenges—Reconceptualizing Security in the 21st Century*, published in December 2007.[4] Prior to the launching of this book at UCS in Washington, D.C. on 21 April 2008, I had my last phone conversation with Jonathan Dean, and in December 2010, after the death of his wife, I received his last handwritten letter.

In August 2013, I approached his long-term close colleague at UCS, Ms. Teri Grimwood, to ask Ambassador Dean's permission to translate his last book chapter into Chinese. In this letter, I asked Teri to consider coediting a volume to honor the achievements of Ambassador Jonathan Dean on the occasion of his 90th birthday in June 2014. Fortunately, Teri agreed. She consulted with Ambassador Dean and his family and both consented to this project. We started work on this book in autumn 2013, when he was already very ill, and completed it in March 2014, 2 months after he had passed away. I am extremely grateful to Teri Grimwood and Alden Meyer at UCS in Washington, who agreed to cooperate on and support this publication.

Dr. Jonathan Dean participated as a keynote speaker at two international conferences organized by AFES-PRESS in December 1989 and in October 1991 in Mosbach in the state of Baden-Württemberg in the southwest of Germany. The sixth International AFES-PRESS Conference on 8–10 December 1989 on "Verification and Arms Control Implications for European Security" took place just 4 weeks after the Berlin Wall had fallen.

During this conference, the Mayor of Mosbach and then Chairman of the AFES-PRESS Board, Fritz Raff, gave a reception in the town hall and a dinner where one of the three speeches on "The Political Tasks of Verification in a European Arms Control Regime" was given by Ambassador Dean. He suggested as part of a regional cooperative security regime for Europe the establishment of Risk Reduction and Crisis Management Centers (Chap. 7).

> According to Ambassador Dean, the required common centers for the coordination of verification tasks and of confidence building should be expanded to deal also with the

[4] Jonathan Dean: "Rethinking Security in the New Century—Return to the Grotean Pattern", in: Hans Günter Brauch, Ursula Oswald Spring, Czeslaw Mesjasz, John Grin, Pal Dunay, Navnita Chadha Behera, Béchir Chourou, Patricia Kameri-Mbote, P.H. Liotta (Eds.): *Globalization and Environmental Challenges: Reconceptualizing Security in the 21st Century* (Berlin—Heidelberg—New York: Springer-Verlag, 2008): 3–„6. This text will be republished in 2014 in Chinese in: Hans Günter Brauch, Úrsula Oswald Spring, Czeslaw Mesjasz, John Grin, Liu Cheng (Eds.): *Globalization and Environmental Challenges: Reconceptualizing Security in the 21st Century* (Nanjing: Nanjing Press Company, 2014).

function of risk control and crisis prevention in Europe. In order to avoid that miscalculations and erroneous assessments would be needed to deal with low-level incidents. Such a common European crisis reduction center could be a starting point of a gradual institutionalization of the CSCE process. Ambassador Dean proposed to expand a single common verification center into a risk reduction center and later on to a fact-finding and mediation center. It should start out with routine activities: notification of force activities, observer and inspection requests, discussion of compliance issues, and defusing low-level incidents.

The abstract of Ambassador Dean's keynote speech is included as text 7 below.[5]

Box 1: "How experts learn to control", in: *Heilbronner Stimme*, 12 December 1989.

Wie Experten das Kontrollieren lernen

Der Mittelstreckenvertrag zur Beseitigung von Pershing und Cruise-Missile und der sowjetischen SS-20-Raketen machte zum erstenmal Überlegungen notwendig, wie in der Praxis der Abbau und die Verschrottung der ausgehandelten Systeme überwacht werden solle. Man einigte sich untér anderem auf Vor-Ort-Inspektionen und ein ausführliches Ritual der gegenseitigen Information vor solch einem Besuch.

Auf diesem Stand sind die beiden Supermächte nicht stehengeblieben. Für den START-Vertrag, der eine Reduzierung der Langstrecken-Atomraketen vorsieht und nach Ansicht der Experten bereits im kommenden Jahr abgeschlossen werden kann, einigten sich nun die USA und die Sowjetunion im Vorfeld auf eine Art bilateralen „Studienkurs".

Dabei geht es um die Untersuchung von Langstreckenbombern, die durch ihr Einsatzkonzept (Atombomben-/Marschflugkörper-Bestückung) unter das START-Abkommen fallen. Die Experten sollen direkt am Objekt lernen, was mit atomaren Marschflugkörpern ausgerüstete Flugzeuge von konventionell bewaffneten unterscheidet. Je zehn Experten besuchen dabei einen Luftwaffenstützpunkt der anderen Seite. Dort erhalten sie vom Militärpersonal eine Einführung besonderer Art: Sie werden anhand der technischen Erläuterungen untersuchen, wo die erkennbaren Unterscheidungsmerkmale der einzelnen Bombertypen liegen. Aus ihren Erfahrungen sollen dann Methoden zur Überwachung entwickelt werden, denn die luftgestützten atomaren Missiles lassen sich nur extrem schwierig von den mit konventionellem Sprengstoff ausgestatteten unterscheiden. -bert

Schlägt für die nahe Zukunft die Gründung einer unabhängigen Kontrollagentur für die europäischen Staaten vor: Jonathan Dean, der frühere Leiter der USA-Delegation bei den Wiener Truppenabbauverhandlungen. Als Aufgaben nannte er die Vermittlung bei kleineren Krisen und die Überprüfung von regionalen Rüstungskontrollverträgen. Foto: HSt-Dirks

After the conference, which had 100 participants from 20 CSCE member countries, we had very intensive conceptual discussions with much good German wine in my private library, involving diplomats from both the US and the USSR, generals, professors, and young scholars, some of whom had been negotiators at the MBFR talks. In these fascinating, private off-the-record and intensive discussions Ambassador Dean was one of the most experienced and active participants in the conceptual debate on the future of Europe at a time when history was still open and many different alternative developments were foreseeable.

Two years later at another historical turning point, when the Soviet Union was dissolving and being replaced by the Russian Federation, Ambassador Dean

[5] See: Brauch (1990)

participated in the 7th AFES-PRESS conference on "Controlling Military Research and Development and Exports of Dual Use Technologies as a Problem of Disarmament and Arms Control Policy in the 1990s". This took place on 24–27 October 1991 in a training center for farmers in Mosbach-Neckarelz, and was attended by 70 participants from 16 CSCE countries and from Bangladesh, India, Japan, the Republic of Korea and Peru.[6]

Box 2: "Romantic," in: *Heilbronner Stimme*, 16 November 1991.

„Romantic"

Jonathan Dean, Ex-Botschafter der USA, mag Mosbach. In bestem Hochdeutsch schwärmt er: „Ich bin begeistert von Mosbach und seinen schönen Häusern." Weil das zu sehr nach Klischee klingt, ergänzt er: „ Ich liebe diese saubere und scharfe Luft hier."

Wenig bemerkt haben die Mosbacher die rund 70 hochkarätigen internationalen Gäste der AFES PRESS Tagung um den einheimischen Politikwissenschaftler Dr. Günther Brauch.

Die Friedens- und Abrüstungsforscher tagen abseits in Neckarelz, graben sich in die brisante Konferenzthematik ein. Da bleibt wenig Zeit für die Umgebung, für Erkundungen des Neckartals oder des Odenwalds.

Der abendliche Gang durchs Fachwerkstädtchen zum Rathausempfang hat Eindrücke hinterlassen. Nicht nur beim Amerikaner Dean. Zwischen Indern, Dänen, Schweden, Briten, Sowjetrussen oder Finnen bewegen sich zwei höfliche Südkoreaner fast erfurchtsvoll im wuchtigen Gewölbe des Rathaus-Untergeschosses.

Zum ersten Mal in Deutschland, des Deutsch unkundig, ringt ihnen das Gesehene respektvolle Worte ab. „Romantic and historic" wirkt auf Il Chul Chung und Kwang Keun Moon das Städtle. Beinahe ungläubig vernehmen sie, daß fast 1000 Jahre Geschichte im Kreuzgewölbe über ihren Köpfen lastet.

Ob sie den Streifzug durchs Mosbach von früher und heute von Bürgermeister-Stellvertreterin Rotraud Laade nachvollziehen, ist den stoischen Mienen nicht abzulesen.

Dr. Brauch strauchelt manchmal beim Übersetzen. Kurfürst, Pest, Sanierung, Mittelzentrum, B 27-Umbau oder Landesgartenschau, das Englisch des Spezialgebietes ist doch eingegrenzt.

Jonathan Dean assistiert, bewegt sich in der deutschen, selbst kurpfälzischen Geschichte wie ein Insider, hilft mit Vokabeln aus.

„Mediäval" ist das häufigste Wort an diesem Abend für die englischsprachigen Gäste. Mediäval ist eben nicht nur das Rathaus. Das Mittelalter hat in Mosbach überdeutliche Spuren hinterlassen. Passend ist der Rahmen fürs Dinner der Stadt, natürlich in einem Gasthaus mit Fachwerkschmuck.

Und genauso zurückgedreht scheint die Zeit beim Kammermusikabend (mit Barockmusik) im gotischen Templer-Haus, einen Steinwurf vom Tagungsort entfernt.

Mosbach, so senkt es sich den Gästen in Erinnerung, scheint Lichtjahre von dem entfernt, was die Konferenz als Problem der Zeit behandelt. maw

6 Brauch (1992)

As the global security adviser of UCS in Washington D.C., he was a frequent speaker both in English and German. When I invited him to speak to my students at the Johann Wolfgang Goethe University in Frankfurt he switched to German and when I had problems during the reception in the city hall in translating a speech into English he helped me out with some technical words, as a local reporter wrote in his commentary "Romantic" (Box 2). In response to my letter congratulating him on his 80th birthday in June 2004 and inviting him to give a keynote speech at our workshop in September 2004 in The Hague, he said that he was unable to attend but he promised to write the first preface to our *Global Environmental and Human Security Handbook for the Anthropocene* (GEHSHA), to be published in Chinese in 2014 (Box 3).

In 1990, when German reunification was approaching, Ambassador Dean wrote an innovative thought piece for our AFES-PRESS Reports on *German Unity and the Future European Order of Peace and Security*, in which he outlined the "Components of a Post–Cold War Security System for Europe",[7] later published in a revised and further developed form in a chapter in the second volume of another three-volume book that I coedited with Robert Kennedy (1990–1992).[8] This conceptual thought piece is included below as Chap. 9.

During the 7th AFES-PRESS conference on 25 October 1991, Ambassador Dean gave a dinner speech suggesting an "Arms Reduction Blueprint" for Europe. This was based on a talk he had given ten days earlier in London (Chap. 10). The paper he presented was later published as "Constraining Technological Weapons Innovation in the Post–Cold War Environment" and is included in this volume as Chap. 11.[9]

This conference and publication was the result of a joint Dutch–German cooperation between the *Inter-university Network for Studies on Technology Assessment in Defense* (INSTEAD) and the INSTEAD Center for Verification Technology headed by Brigadier General Henny van der Graaf (Retired). The German Research Foundation (DFG) funded the conference and the Free University of Amsterdam covered the costs of the preparation and publication of the resulting book.

During the AFES-PRESS business meeting during this international conference, which took place in October 1991, the board of AFES-PRESS invited Ambassador Dean to succeed Mayor Fritz Raff as chair of the Scientific Advisory Board of AFES-PRESS, an honorary position he held until his death. From 2007 he was cochair with Prof. Dr. Ursula Oswald Spring, the first female minister of ecological development in the Mexican state of Morelos. The AFES-PRESS board will remember the many conceptual discussions with Ambassador Dean and his intellectual contributions that were often ahead of official policies. He will be remembered as a delightful colleague and as a fatherly friend.

---

7 See: Afheldt et. al. (1990)

8 See: Hans and Kennedy (1993)

9 Brauch et. al. (1992)

Box 3: My letter congratulating Ambassador Jonathan Dean to his 80th birthday.

***P*eace *R*esearch and *E*uropean *S*ecurity *S*tudies (AFES-PRESS)**

PD Dr. Hans Günter Brauch, Alte Bergsteige 47, 74821 Mosbach, Germany, 1 June 2004

☎ 49-6261-12912 49-6261-15695 brauch@afes-press.de http://www.afes-press.de

Amb. Jonathan Dean
10020 Colvin Run Road
Great Falls, VA 22066
USA
Email. jdean@ucsusa.org

Re: Announcement of the *First AFES-PRESS Grotian Award to Amb. Jonathan Dean*

Dear Amb. Dean, dear fatherly friend,

It is great pleasure and honour for me to inform you about the decision of the AFES-PRESS board to honour you for your lifelong activity for enhancing transatlantic cooperation and German-American reconciliation with our first *AFES-PRESS Grotian Award* during the opening session on the eve of the 5th Pan-European Conference on International Relations in The Hague in the Netherlands that will be hosted by a representative of the Dutch government.

We would like to invite you to attend our AFES-PRESS workshop on Reconceptualising Security in an Era of Globalisation and to give a speech during the opening session on 8 September 2004 where we plan to honour you in person with our first AFES-PRESS Grotian Award. We would like to invite you to share with us in your acceptance speech your thoughts on:

Rethinking and Reconceptualising of Security:
A Grotian perspective on the need for a new transatlantic consensus

AFES-PRESS is pleased that we are able to reimburse your full travel costs from Washington to Amsterdam. During your stay in The Hague you will be the guest of AFES-PRESS. This trip was made possible by a grant we have received from NATO. We would also like to publish a revised version of your speech as a preface essay in our next major reference book to be published by Springer in 2006 in our Hexagon Series to be edited by:

Hans Günter Brauch, Ursula Oswald Spring, John Grin, Czeslaw Mesjasz,
Navnita Chadha Behera, Béchir Chourou (Eds.):
Facing Global Environmental Change and Globalisation
Reconceptualising Security in the 21st Century

We also send you our warm and best wishes for your 80th birthday in June 2004 for good health, personal happiness and continuing good cooperation in the years to come.

With kind regards

on behalf of the AFES-PRESS board

Yours faithfully

Hans Günter Brauch

Box 4: Jonathan Dean's personal letter to the author on the eve of his 80th birthday, reproduced here with the permission of Ambassador Dean's daughter, Theodora Schiro.

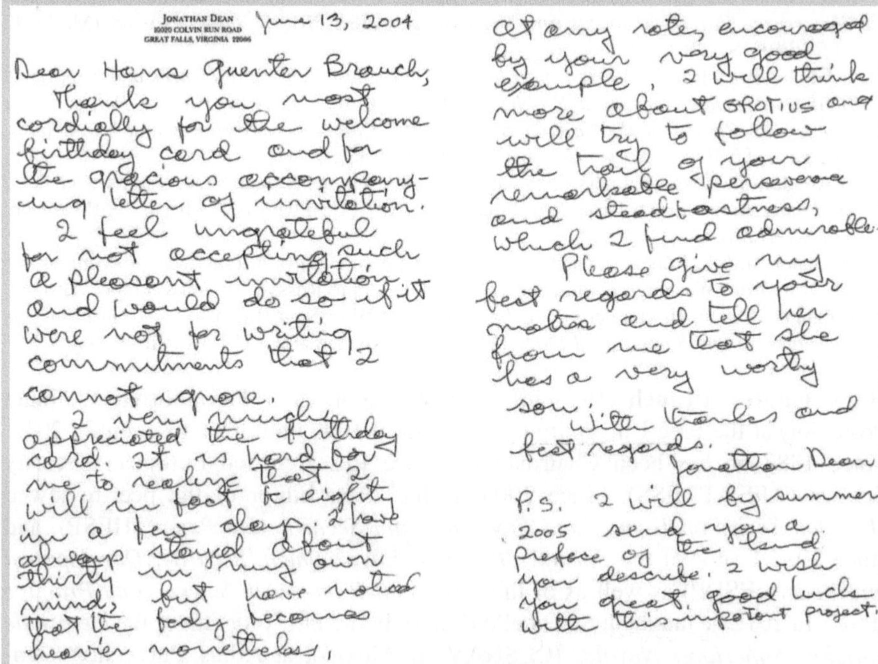

JONATHAN DEAN
10020 COLVIN RUN ROAD
GREAT FALLS, VIRGINIA 22066

June 13, 2004

Dear Hans Guenter Brauch,

Thanks you most cordially for the welcome birthday card and for the gracious accompanying letter of invitation. I feel ungrateful for not accepting such a pleasant invitation and would do so if it were not for writing commitments that I cannot ignore.

I very mighty appreciated the birthday card. It is hard for me to realize that I will in fact be eighty in a few days. I have always stayed about thirty in my own mind, but have noticed that the body becomes heavier nonetheless, at any rate, encouraged by your very good example, I will think more about GROTIUS and will try to follow the trail of your remarkable perseverance and steadfastness, which I find admirable.

Please give my best regards to your mother and tell her from me that she has a very worthy son.

With thanks and best regards.

Jonathan Dean

P.S. I will by summer 2005 send you a preface of the kind you describe. I wish you great good luck with this important project.

This volume is a true transatlantic and American–German cooperation between Teri Grimwood of the *Union of Concerned Scientists* (UCS), one of the largest American scientific NGOs, and AFES-PRESS, a small international scientific society under German law.

In short, Ambassador Jonathan Dean has been a distinguished American diplomat, a concerned scientist, a policy-oriented, forward-looking, innovative thinker, a global citizen, a friend of Germany, and a fatherly friend to me. I have always valued his professional and personal advice highly during the long period he was chair and later cochair of the Scientific Advisory Committee of Peace Research and European Security Studies, a nonprofit international scientific society under German law, from 1991 until his death on 14 January 2014.

Mosbach, Germany, March 2014 Hans Günter Brauch

References

Hans Günter Brauch 1990: "Verification and Arms Control—Implications for European Security: Main topics and Results", in: Hans Günter Brauch (Ed.): *Verification and Arms Control—Implications for European Security—The Results of the Sixth International AFES-PRESS Conference—Part I: Abstracts and Discussions*, AFES-PRESS Report No. 35 (Mosbach: AFES-PRESS): 40.

Hans Günter Brauch (Ed.) 1992: *Controlling Military Research, Development and Exports of Dual Use Technologies as a Problem of Disarmament and Arms Control Policy in the 1990s—The Results of the Seventh AFES-PRESS Conference, Abstracts and Discussions*, AFES-PRESS Report No. 45 (Mosbach: AFES-PRESS).

Horst Afheldt, Malcolm Chalmers, Jonathan Dean, Hans Günter Brauch 1990: *German Unity and the Future European Order of Peace and Security*, AFES-PRESS Report No. 38 (Mosbach: AFES-PRESS).

Hans Günter Brauch, Robert Kennedy (Eds.), 1993: *Force Posture Alternatives for Europe After the Cold War* (Washington–Philadelphia–London: Crane Russak, Taylor & Francis Group).

Hans Günter Brauch, H. van der Graaf, John Grin and Wim Smit (Eds.), 1992: *Controlling the Development and Spread of Military Technology* (Amsterdam: VU University Press).

Hans Günter Brauch (Dr. phil. habil.) taught as a Privatdozent (Adjunct Professor) at the Free University of Berlin from 1999 until his retirement in 2012. Since 1987 he has been Chairman of Peace Research and European Security Studies (AFES-PRESS). Since 2003 he has been Editor of the peer-reviewed *Hexagon Series on Human and Environmental Security and Peace* (HESP), and since 2012 Editor of the *Springer Briefs in Environment, Security, Development and Peace* (ESDP) as well as of this *Series on Pioneers in Science and Practice* (PSP). In 2012 he taught at SciencePo (Paris), in the PhD program of the *Centro de Estudios Superiores Navales* (CESNAV) in Mexico, at Arhus University (Denmark), at the *European Peace University* (EPU) (Austria), at the *National University of Malaysia* (UKM), and at Chulalongkorn University (Bangkok); and from October 2013 until January 2014 he was visiting professor at the Faculty of Political Science of Chulalongkorn University (Bangkok), Thailand.

Ambassador Jonathan Dean. *Source* Personal photo collection of Ms. Theodora Schiro, daughter of Jonathan Dean. Reprinted with permission

Contents

Part I On Ambassador Jonathan Dean

Part II Texts by Jonathan Dean on Detente and Arms Control in Europe in the 1980s

Part I
On Ambassador Jonathan Dean

Ambassador Jonathan Dean relaxing and playing the accordion in his home. *Source* Personal photo collection of Ms. Theodora Schiro, daughter of Jonathan Dean. Reprinted with her permission

Chapter 1
Biography Jonathan Dean

Jonathan Dean was born on June 15, 1924 in New York City.[1] He was the oldest of three children, and was raised in Darien, Connecticut until the death of his mother early in his life. His father then moved the family back to New York City, where Dean attended the Riverdale Country School before entering Harvard College at the age of 16.

1.1 Early Years

In December of 1941 after the attack on Pearl Harbor he joined the Canadian Army while visiting relatives in Canada. At 17 he was too young to join the American Army without parental permission, and knowing that he would not receive it took another route into the war, as did a good number of young American men.

After training in Vancouver, his unit was transferred to Britain in late 1943. He was assigned to a Field Security Unit, the British-Canadian equivalent of the American *Counter Intelligence Corps* (CIC). He served in it until the invasion, landing 2 days after D-Day. About a month after the invasion, while in Normandy,

[1] This biography is extracted from a multi-day interview of Ambassador Jonathan Dean by Charles Stuart Kennedy of the Association for Diplomatic Studies and Training, Foreign Affairs Oral History Project. The initial interview date was July 8, 1997, and the final interview date was November 10, 1998. The interview is at: http://www.adst.org/OH%20TOCs/Dean,%20Jonathan.toc.pdf. The permission to use this material was granted on 21 March 2014 by Ms. Marilyn Bentley, Business Manager, CATO Institute on behalf of the Association for Diplomatic Studies and Training; See also the website on this book, at: <http://www.afes-press-books.de/html/SpringerBriefs_PSP_Dean.htm>.

H. G. Brauch and T. Grimwood (eds.), *Jonathan Dean*,
SpringerBriefs on Pioneers in Science and Practice 19,
DOI: 10.1007/978-3-319-06662-2_1, © The Author(s) 2014

he was transferred back to the U.S. Army, the 407th Infantry Regiment of the 102nd Infantry Division of the Ninth Army.

While crossing the Ruhr River, a tributary of the Rhine near Bielefeld, Dean was wounded. He was sent to hospital in Paris and then went back to his unit, which by that time was mechanized and moved on to the Elbe. It met the Soviet forces near Tangermünde-Tangerhütte. The unit then withdrew through Thuringia in late 1945 to the border of Bavaria, where he stayed a short time then returned to the United States and was mustered out.

Dean then returned to Harvard College, but transferred to Columbia College to finish his degree. There, at the start of the Cold War, he signed up with the Columbia Institute of International Affairs, graduating in 1948. While at Columbia, he had contact with professors who stressed the importance of the United Nations.

1.2 Starting His Diplomatic Career in Germany (1950–1956)

1.2.1 First Assignment as a Kreis Officer

In 1949 he took the Foreign Service exam, and went on the waiting list for Foreign Service officers. In 1950 he married Theodora Dean, a neighbor introduced by his father from the family's early days in Darien, Connecticut. The Department of State, having taken over the occupation of Germany from the Army in an effort to civilianize the process, took people from the waiting list and gave them staff appointments as Kreis Officers in West Germany as representatives of the High Commission. His assignment was in Limburg on the Lahn, which is the seat of the bishopric of Limburg. He was also responsible for the adjoining Oberlahn Kreis.

Father Kenneth Dean, Jonathan Dean, Brother David Dean at Jonathan Dean's wedding, March 4, 1950. *Source and copyright* Private photo collection of the Dean family

J. Dean and friends at his wedding, March 4, 1950. *Source* Private photo collection of the Dean family

This was a residual function of the U.S. occupation of Germany. In the spring of 1950, the High Commission was actively engaged in the effort to democratize post-war Germany. That was also then the main function of the Kreis Resident Officers—it was an educational experience on the fundamentals of democracy. A great deal was achieved in the direction of democratization, voter education, formation of free trade unions, free press and free radio.

1.2.2 The Rearming of Germany

Jonathan Dean's next assignment was in the High Commission in Frankfurt, where part of the U.S. High Commission still remained during the move to Bonn. His job was to follow neo-Nazi developments, to analyze reports, and to pull them together. He did that for about three or four months, and then was shipped up to Bonn in 1951 to help organize the new German army. At this time he was a regular Foreign Service Officer, and remained in Bonn until 1956.

He went to Bonn following neo-Nazi extremism, and ended up in the political section. Dean was one of the reporting officers, and shortly was assigned to the Liaison House, a contact place with the new German government and Theodor Blank, the German trade union leader who was chosen to set up the new democratic German armed forces, and his Amt Blank. The Liaison House worked mainly with the parliament, the chancellor's office, and with the Amt Blank, as it developed with the new German army. The effort was to develop a new form of discipline, based more on teamwork and less on the hierarchical system of the old Wehrmacht. Dean took the first group of deputies from the Bundestag Defense Committee to the United States in 1955 to look at how it was done in the United States as a prelude to their own defense legislation. They passed good legislation, including a law providing for a Wehrbeauftragter of the Bundestag, a kind of ombudsman who is supposed to ensure the democratic conditions inside the armed forces.

Jonathan and Theodora Dean, Limbug (Germany), 1950. *Source* Private photo collection of the Dean family. Permission was granted

Dean stayed in Bonn for the next 3–4 years, working on the military buildup and doing the domestic political reporting about all of the treaties that brought Germany into NATO and formally ended the occupation. From the outset the Social Democrats had been very strong opponents of German rearmament. They thought it was socially wrong, politically wrong, morally wrong. They finally reconciled themselves to German membership in NATO and accepted it as desirable.

Jonathan Dean with a delegation of the Defense Committee of the German Federal Parliament (Bundestag) visiting San Francisco in 1952. Among the Members of Parliament are (from *right* to *left*) Fritz Erler (SPD), Dr. Richard Jaeger (CSU), Chairman of Bundestag Defense Committee, General Hasso-Eccard Freiherr von Manteuffel (FDP) and a representative of the German Defense Ministry

1.3 State Department in Washington, D.C. (1956–1960)

1.3.1 The Hungarian Uprising

In 1956 Dean returned to Washington, and was assigned to the East German desk in the State Department for the next 2 years. He was totally shaken by the inaction of the United States at the time of the Hungarian uprising, and suggested that the outcome of Hungary demonstrated that the Warsaw Pact system was going to be there for a long time so it was necessary to become more insidious and more refined. He came to the conclusion that the Soviet and communist system was now relatively firmly established in Eastern Europe and that the way to deal with it was with what was then called a policy of engagement. In 1960, he was working on the Paris Summit when Gary Powers' U-2 airplane was shot down, ending the opportunity. At this time he applied for Czech language training and was assigned to Czechoslovakia in 1960.

1.4 At the U.S. Embassy in Prague (1959–1962)

From 1959 to 1962 Dean was first secretary at the U.S. Embassy in Prague for economic and political issues. Czechoslovakia was the last holdout of Stalinism and they even had an intact statue of Stalin. Anton Novotny, the party head, was an old line Stalinist and did not permit even a slight movement or change. That spring when Dean arrived they had a traditional May Day celebration in which people recited a poem called Queten, or May. This was certainly not an active resistance but they were able to get six or eight students together for this purpose. The regime was so thoroughly oppressive that they really couldn't get off the ground. Yet it was these celebrations that in the Dubcek era became the focus of the Prague Spring.

At this time it was continually feared that U.S. prestige and our support for West Germany would be placed into question by some Soviet blockage of the access to Berlin. The problem was that, in the immediate post-war period, agreements were reached about military access to West Berlin for British, French and American troops. There was no agreement about access of German civilians. It was assumed that movement from all parts of Germany would be at a low level. When West Germany was separated from East Germany, then access to and from Berlin, civilian access as well as visits to East Berlin by both West Berliners with relatives there and West Germans with relatives there, became a very important issue.

The East German government was seen as a creature of the Soviet system but also as a people that wanted to build up their own status. Their intelligence activities in West Germany were beginning to be revealed and of course they were enormous, extending into the chancellor's office at the time of Willy Brandt.

At this time the migration of East Germans into West Germany was becoming a major issue. East Germany was clearly losing its intellectual cream and of course West Germany was receiving the refugees, taking care of them, and actually subsidizing them. That was a condition which actually continued right up until nearly the time of unification. It was a good deal for these individuals because they received economic help from the West German government, but it did place in question the success of the system, not its hold over the situation but its economic and social success, and that definitely was open to question.

1.5 Elizabethville, Belgian Congo (1962–1966)

Dean was suddenly transferred in 1962 to Elisabethville in the Katanga, Congo. The only preparation he had was the riots after the death of Lumumba that had occurred in Prague.

Belgian government policy and American policy was completely disunified and operated by different groups. The Social Democratic Belgian government was completely split. The industrial mining interests, Union Miniere, favored Tshombe, who was their creation. The Belgian government of Paul Henri Spaak supported the central government in Kinshasa in this situation where Katanga had seceded and declared its independence.

There had been one conflict between the United Nations forces, which were in effect surrounded and beleaguered in Elisabethville, and the Katangan Gendarmerie. The latter were the remnants of the earlier Congolese armed forces—the Force Publique—taken over by the Katangans. This force had retained some Belgian officers, and were much better disciplined, better paid than Congolese troops in the rest of the Congo. They were stiffened by quite a large contingent of European mercenaries.

The UN forces were composed mainly of an Indian brigade, surrounded and penned up in the city. Subsequent to that a systematic program of insults to the UN forces took place. The Katangan equivalent of Nigerian market women beat up Indian UN soldiers and things of that kind. The Indian army had been quite humiliated by the performance and put in troops with combat experience in the recent border war with China.

Two days after Dean's arrival he was summoned to a meeting with Tshombe who launched into a tirade about what the UN, backed by the U.S., had done to Katangan soldiers. At the end of this harangue, he pulled a blanket off two stretchers showing two very badly burned bodies. His claim was that the UN had dropped napalm furnished by the U.S. on a Gendarmerie position. Dean said he didn't know anything but would look into it.

The next day, Dean went out to the place where this action had taken place. It was rather clear that the bodies had been prepared by someone assisting Tshombe to make a television case; this was broadcast to Europe and to the United States. The U.S. embassy was strongly supporting the official UN policy, which was to try to defeat the secession and to reunite the Congo. The U.S. fear was that if the Congo was 'balkanized,' the term used at the time, it would become a sort of shifting, unstable center for the whole of Africa and create impossible political and economic conditions. At the time that Lumumba was deposed, several score Warsaw Pact advisors were arriving every day to take over various functions of the Congolese government.

Little or no progress was being made in resolving the secession. Dean gradually became aware of the fact that the UN force seemed to be preparing for further conflict, and reported this to Washington. The denouement came at Christmas 1963 when the UN forces carried out a very successful, relatively bloodless operation with few casualties among the Gendarmerie and still fewer among the UN forces. The issue was whether they should proceed in hot pursuit. Dean told them that they should go ahead and do it. This did end the secession. It reunified the Congo and brought back the foreign exchange from the copper exports of Katanga. Of course the Congo's troubles were not over by any means.

In 1964 Jonathan Dean returned to the United States and attended the National War College.

1.6 National War College (1964–1965)

From 1964 to 1965 Jonathan Dean attended the National War College, which was totally absorbed in the Vietnam issue. At this time Dean objected to the ground force involvement in Vietnam, while recognizing that national prestige had been engaged and it would be difficult to get out without a victory of some kind. On the basis of his Congo experience, he felt Vietnam lacked the people infrastructure for large-scale involvement. At this point Dean felt there was less feeling about the military winability, but concern about whether it was justified politically.

1.7 United Nations, International Organizations (1965–1966)

In 1965 Dean went to the United Nations, deputy director of the office of UN Political Affairs which works with the Security Council. He was assigned to the International Operations (IO) Bureau because of his engagement in the peacekeeping exercise in the Congo. It was not his first choice. He had been in line for assignment to Warsaw as Deputy Chief of Mission, but that fell through. At that time the IO Bureau and the United Nations were not held in high esteem in the State Department. The bureau was regarded as an unserious operation which continually interfered in an amateurish way with the more solid operation of bilateral-lateral relationships.

Dean worked on Kashmir and in particular on Rhodesian sanctions. That was the time when Ian Smith had seized power in Rhodesia, leading to sanctions against Rhodesia. This was the trial run for what later happened with South Africa, where a set of sanctions was indeed effective in persuading the South African business community that they had to end apartheid.

At this time the second Indian-Pakistani war over Kashmir had occurred and there was a peacekeeping mission of the UN there, an observer mission type. To get it there and to get the various UN resolutions backing it was the main work for quite some time.

While in IO, he encountered serious people who were dedicated to the concept of the United Nations. He was later to say, "The United Nations is really a tiny organization which carries on its shoulders the hopes of most of mankind. Consequently, it either infuses you with contempt or inspiration."

1.8 Assistant to Robert Bowie (1966–1968)

Robert Bowie was on his second tour of duty as Counselor to the U.S. State Department. In his earlier tour he had been Counselor and Head of the policy planning sector. Then it was the Cold War. This time it was Vietnam, which overshadowed everything, and Robert Bowie had an uneasy relationship with Secretary Rusk because of Bowie's skepticism about Vietnam, causing him to be underworked.

1.9 Bonn and the Berlin Quadripartite Agreement (1968–1972)

In 1968 Jonathan Dean was assigned to the U.S. Embassy in Bonn, 2 weeks before the actual Soviet invasion of Czechoslovakia. Receiving a phone call from the EUCOM commander, who was about to launch a very big military exercise along the Czechoslovak border using both German and American military forces, he was asked whether he should go ahead with the exercise. Dean told him he should suspend it immediately and he did so. There was some criticism afterward that it telegraphed the fact that we weren't going to interfere and this made it easier for the Soviets to carry out this invasion. Dean felt that it was the absolutely right thing to do under the circumstances.

It did confirm the ongoing Cold War and the need for an American military presence in Europe. But it didn't hold down the German thinking on the need for change from the Cold War line. It did answer critics in the United States of the American troop presence in Germany.

Dean's first interview in the Foreign Ministry was with Ulrich Sahm, who was the head of West Germany's political department. He was told of the German interest in something being done to deal with the short-range Soviet nuclear weapons stationed in Eastern Europe. The SALT talks were supposed to begin and had to be postponed by President Johnson because of the Soviet invasion of Czechoslovakia. The Germans felt that their security concerns were not taken into consideration in the preparations to negotiate the reduction of strategic weapons aimed at the United States. This concern went on and on and played a role in German thinking, until it later culminated in the great missile confrontation with the Pershing IIs and the SS-20s in the early 1980s.

Can you really trust another country with final responsibility for your own security? It was a very deep problem and remained such throughout the Cold War. It was a recurrent theme in German-American relations together with the other theme already mentioned—German neutralism or whether Germany would be a reliable ally or not. This time the distrust was on the side of the United States. These two factors—U.S. nuclear protection and German unity—were the two main factors in the German-American relationship. German unification was always believed by Americans to be the lever by which the Soviet Union could affect the loyalty of Germany and bring it over on its side. The theme of German Ostpolitik came up in full force after Czechoslovakia, which many Germans regarded as the last paroxysm of Soviet imperialism.

Willy Brandt was the German foreign minister at that time and in December of 1968 he raised the issue in NATO. NATO foreign ministers having responsibility for Germany, that is the United States, Britain and France, together with Germany, formed the so-called Bonn Group in Bonn, usually staffed by the political counselors, unless a high-level issue was involved. The Bonn Group foreign ministers, the four-powers, met on the edge of the annual NATO meetings and reviewed the situation.

In that meeting, Brandt raised the possibility of negotiating on Berlin. His chief assistant, Egon Bahr, had extensive discussions with a Soviet representative in East Berlin about this issue and they apparently had indicated that they were willing to make a deal on Berlin. Anyhow, Brandt's proposal gestated for another year or so before it really became a prominent issue.

The Bonn Group on the operating level consisted of the political counselors of the British, French and U.S. embassies and the deputy political director of the Auswärtiges Amt, most of the time Gunther van Well, who was later German ambassador to the UN and to Washington. He was a very gifted interpreter of German policy. This man kept the allies more or less satisfied during the most intense and revolutionary period of German Ostpolitik by giving them his interpretation, his version, of what was going on and explaining to them each move in advance.

At this time Henry Cabot Lodge was ambassador in Bonn, and after him, Kenneth Rush.

The stream of reports that the Bonn Group sent in more or less satisfied the home foreign ministries. Henry Kissinger was very worried about the Germans getting out in front and taking the lead, the initiative, in dealings with the Soviet Union and undermining the authority of the United States. He indicated at various times his doubts and dissatisfaction.

Dean advocated various versions of the policy of engagement with Eastern Europe. He considered this the right time to try this approach in East Germany and Eastern Europe, feeling it would contribute to stability in East–West relations, satisfy the German desire to play a more important role in their own future, a role which they merited, and would actually exploit an excellent tactical opportunity because the Soviet Union was making such a strong effort to reestablish themselves in world opinion after the invasion of Czechoslovakia. Dean thought that the Soviets could be brought to pay something in terms of actual concessions, especially on Berlin, the main area of our own security concerns and risks.

The Soviets had alienated all of the Warsaw Pact countries including Romania. Fear of Germany had served as the only real glue there was for the Warsaw Pact, and here was Germany acting in a somewhat civilized manner. Later on, of course, as part of the Ostpolitik, Germany made great efforts to regularize its relationships with Poland, Czechoslovakia, and all of the Eastern European countries. These efforts paid off in 1989–1990 as the Warsaw Pact system collapsed in Eastern Europe.

Ambassador Kenneth Rush was vigorous and determined. He had been Nixon's law professor at Duke and Nixon asked him later on to take on his defense in Watergate. Rush arranged that the president be told by various other officials in the administration that this would be a terrible mistake because Rush was so involved in serious foreign affairs at that time that he couldn't be spared. He was a far more vigorous and far more take-charge guy than Ambassador Lodge.

Tentative preparations had already begun for the Berlin negotiations and Ambassador Rush really took it up with great vigor. He had established a back channel communication with Valentin Falin, the Soviet ambassador in Bonn, and with the very same Egon Bahr who was state secretary in the chancellor's office. Bahr was Brandt's chief assistant for foreign affairs and particularly for relations with the East.

Willy Brandt, who originated this on the basis of Bahr's talks with the Russians, told the foreign ministers in December of 1968 in Brussels at their Bonn Group meeting that he thought it would be possible to negotiate with the Russians on access to Berlin. This was the neglected area in wartime agreements having to do with setting up the four power control in Berlin and Germany. What had been neglected was German civilian access. All the arrangements had to do with military access and somehow it was assumed that civilian access would be automatic or that there would be only limited travel by German civilians. Of course, the division of the country was not foreseen at that time during those '44–'45 negotiations in London about the occupation zone, the command authority, control of Berlin, and so on.

Anyhow, Brandt insisted that it would be possible to improve the situation somewhat, obtaining more passes for West Berliners to visit East Berlin. The people operating the western mechanism in Berlin, that is, the commandant who was a military officer but also the State Department officers, had extremely strong possessive feelings about Berlin. Beyond that, they thought the mechanism that had grown up in and around Berlin was so intricate and so delicate that efforts to open it up in negotiation would very probably result in its breakdown, so they very strongly opposed these negotiations.

Henry Kissinger didn't like these Berlin negotiations either because they represented in his eyes an excessive German independence in relations with the Soviet Union and a dangerous area. Right from the division of Germany, the idea that Russia could neutralize the whole of Germany by propaganda, by seemingly serious proposals to reunite the country, were combined with a low estimation of German common sense and willingness to withstand these appeals to national feeling. The U.S. fear from the outset was that the entire Cold War could be lost with some brilliant Soviet propaganda which Western democracies, being subject to democratic controls, could never hope to equal and that the Soviets would in this way achieve political and military victory. In fact, when Europe saw any German discussion of reunification and so forth, this development was viewed with intense suspicion by successive administrations in Washington, probably at its most intense with Kissinger. This U.S. concern was generic, deep-seated and very widespread. This is one of the prices Germans had to pay for Hitler and the war. Of course Germany continued to be divided and this concern continued right up through Willy Brandt and beyond.

Later, it became known to Kissinger and the administration from intelligence sources that Egon Bahr, Brandt's chief assistant in Berlin and then in Bonn, was continuing to see a Russian KGB man in East Berlin. This confirmed strong suspicions of the un-wisdom of the German Ostpolitik. There were lots of questions from the press back here and some in Bonn about whether we were losing confidence in the Germans and the U.S. embassy fended those questions off.

It did turn out that this KGB contact had practical consequences, preparing the entire sequence of German treaties with East Germany as well as parts of the Berlin agreement, which we then moved to negotiate on. This Bahr contact was picked up by the Russian ambassador in Bonn, Falin, with Ambassador Rush and did facilitate this Berlin negotiation, the Berlin quadripartite agreement.

A great deal was changing in Germany. Fears of the Cold War were gradually dissipating with a more positive relationship with the Soviet Union and with East Germany. The status of women, in particular, was becoming much more active. The young Socialists with Karsten Voigt were organizing all kinds of rallies and so forth which 10, 15 years later culminated in immense anti-nuclear demonstrations in Bonn which were equaled in the United States by a rally in Central Park with a million people on essentially the same thing. That was the time of the real nuclear controversy, which was yet to come at that point.

Another thing that was important at that time in that period in Germany was the development of a critical press and public media. This came out more clearly later, but even at that time the German press was critical of the government no matter which party was in power. This was most extreme in the case of the *Spiegel* magazine, with never a positive word for good performance. Many of the people in the U.S. Embassy felt that this practice could critically undermine public respect for the new institutions of Germany and that it was not good. But Germans were able to digest this negativism together with these huge long reports of all kinds of facts, some of them rather skewed.

Television, too, was very critical towards the German government, both the Social Democrats and later the Christian Democrats. They were criticized with extreme vigor and this was one more, positive sign that the German democracy had passed the test. In many ways, Germany today is a more thoroughly democratic society than the United States.

During this time President Nixon paid a visit to Europe, his first visit being Germany. He was accompanied by Kissinger, and the administration took over the whole embassy. Dean was in charge of the arrangements for that visit but then committed an error. He was supposed to bring Kissinger and Franz Joseph Strauss together but sent one to one address and one to another. At this point he was removed from the task.

Nixon was well respected in Germany, the imperial presidency. John Ehrlichman, the administration's advance man, got quite a large part the embassy together and gave a big talk on what the president wanted, what he expected, the atmosphere, the product that was supposed to come out of this trip, and so forth. Today that kind of advance team (three or four others, Haldeman was also there) is not viewed as anything exceptional. But it was the inauguration of that kind of megalithic style that we have maintained since. This was the imperial presidency and the visit was a successful visit.

At this time Henry Cabot Lodge was ambassador. He was a very affable, genial man. He had been a journalist and therefore had an intense distrust of journalists.

He and his wife were well liked by German leaders. He was obviously an aristocratic type of person, but he was believed to have the ear of the president, so Germans were well satisfied.

Kenneth Rush came in as ambassador, and was far more vigorous, far more determined. Dean became at this time deeply involved in negotiations on the Berlin treaty. Dean was convinced that Soviet efforts to reestablish the foreign standing of the Soviet Union after the invasion of Czechoslovakia could be exploited and that the Soviets could be made to pay with very specific improvements and demonstrations of these alleged good intentions. He also believed that it would be a mistake to try to frustration, or to be seen to frustrate, the Germans in the pursuit of what was after all their most intensely felt national interest. Even on those grounds alone, even if it didn't work, he felt that the U.S. should support German policy.

Ultimately it was decided first by the four governments and then backed by.

NATO, that the Berlin talks should take place. The ambassadors of the four wartime allies, the three western ambassadors and Soviet Ambassador Abrasimov were going to be the negotiators. Since the Bonn Group teams, the one dealing with Berlin access, Berlin issues, all-German issues, were right there with the Western ambassadors at their embassies, the three political counselors were in effect co-opted to be the negotiators and deputies for the ambassadors. This outcome was received with some annoyance by our colleagues in Berlin and that tension continued all the way through the talks right up to the end.

Henry Kissinger was clearly key in determining policy at this time, and didn't like these Berlin negotiations because they represented in his eyes an excessive German independence in relations with the Soviet Union. In fact when Europe saw any German brewings, or even any German discussion, of reunification, or conditions for reunification, and so forth, this development was viewed with intense suspicion by successive administrations in Washington and probably maybe at its most intense with Kissinger. That kind of manipulation would surely have appealed to him. Of course Germany continued to be divided and this fear, and this concern continued of course right up through Willy Brandt and beyond.

At a certain point in the developments the Soviets stopped making efforts to suggest all-German elections, and concentrated instead on bolstering East Germany and the whole Warsaw Pact structure and on consolidating what they had. That was the threat behind Khrushchev's renewed blockade of Berlin in '61' when he threatened to transfer to the East Germans the complete responsibility for maintaining Berlin and access to Berlin. It was feared then that the East Germans would be far less responsible and far more inclined to extreme behavior than the Soviets because they did not have, shall we say, the worldwide interest in preventing an all-out conflict with the United States that the Soviet Union did. That was the background for these discussions about Berlin access.

There was a fear in Washington that East Germany would end up with its status increased. Indeed, that was exactly what did happen in that East Germany in practice became associated with the four wartime allies and the Federal Republic of Germany in the administration of Berlin. That was the practical outcome of these Berlin negotiations. And, of course, as part of the whole complex, the

Federal Republic recognized East Germany as an independent state. They assumed diplomatic relations with one another and both German nations entered the United Nations as independent states. What some had feared as the outcome was indeed the outcome, but Dean felt it was constructive.

The desire of the German people in both parts of Germany was indeed for unification and that played a role. There was fear of German military intervention. But German forces were so entwined in NATO that it was not a serious possibility. Of course the basic importance of Ostpolitik was that they convinced the East German officials that, if there was some kind of political settlement, they were not going to be taken out and executed by the victors. That eliminated probably the most basic reason for resisting the later unification.

The basic deal in the quadripartite agreement was that the Soviet Union explicitly accepted continuing responsibility for access to Berlin, including specifically civilian access which had been omitted in the wartime agreements. It thus asserted its preeminence over the German Democratic Republic which was making continuing moves to take over that responsibility and pushing the Soviets to get out of business. The latter was an outcome feared by Western countries because it was felt that the GDR had less responsibility and more parochial interests—and the U.S. had fewer points of leverage over it.

Second, in return, the Soviet Union was given an assurance that the Western sectors of Berlin, for which the three Western allies were responsible, were not a part of the Federal Republic of Germany. It was obvious that in the longer run, the Federal Republic of Germany was going to take increasing responsibility for the western sectors and logically should incorporate them into the Federal Republic. The Soviet Union wished to prevent that. In return for stating that the Western sectors did not belong to the Federal Republic, Soviet agreement to the idea that the relationship between the Federal Republic and the Western sectors would be intensified. The third thing agreed to by the Soviet Union was what turned out to be a much more generous regime for giving passes to visit East Berlin and East Germany for West Berliners who had been then cut off largely for many years. The fourth was untrammeled road and rail access to West Berlin for people from the Federal Republic.

1.10 The Mutual and Balanced Force Reduction Talks (1972–1981)

Dean returned to Washington where Kenneth Rush was deputy secretary of State. With Rush's recommendation, he was assigned to the NATO-Warsaw Pact force reduction talks that were just getting under way. Rush designated Dean as the official in charge of the preparations for the negotiations on *Mutual and Balanced Force Reductions* (MBFR) between NATO and the Warsaw Pact. Dean was interested in doing this because he felt this was the military component of ending the Cold War, wherein the Berlin agreement had been the first political part.

Dean was involved in MBFR from 1972 to 1981, most of which was conducted in Vienna. The first part of 1972 was taken up with negotiations among the Washington agencies, then with the NATO countries, and then preparatory talks with the Warsaw Pact in Vienna. The MBFR negotiations started in the fall of 1972.

Jonathan Dean in Vienna during the MBFR talks (1973–1982). *Source and copyright* Private photo collection of the Dean family

The talks were encouraged by a statement by Brezhnev in Tbilisi that, in this issue of force reductions we "should taste the wine." That encouraged the Western countries to move ahead. The Germans were quite intent that if there was going to be a military detente or any form of disarmament between the United States and Russia on the strategic level, they too should benefit and that there should be also a conventional force reduction. They were the chief instigators on the NATO side, together with the smaller NATO countries, Netherlands and Belgium. They wanted force reductions and they wanted to be part of detente if it was going to take place. Donald Rumsfeld was the U.S. NATO representative at that time.

Jonathan Dean with the U.S. MBFR Delegation, Vienna, Austria, April, 1974. *Source and copyright* Private photo collection of the Dean family

Dean went to NATO to present preliminary Washington views on the Western negotiating position for discussion in the Alliance. The U.S. had tentatively agreed with the Soviet Union that negotiations could take place in Vienna. Maintaining NATO coherence and unity in this first negotiation was an extremely important value, probably higher in the minds of many of the governments, including the U.S. government, than making any progress on these negotiations.

The cohesion issue arose at the very outset. Yuri Kwitzinski, later the Soviet ambassador in Bonn and the acting foreign minister at the time of the anti-Gorbachev coup, and a member of the central committee, had been Dean's opposite number in the Berlin talks.

For months in the MBFR talks, a plenary session was prevented because of another issue of controversy between NATO and the Warsaw Pact. The NATO Council had first decided that Hungary should not be among the countries whose own domestic military forces should be reduced as well as Soviet forces and then had reversed itself and said Hungary should be included. The Soviets in Hungary were not forces intended for the central front if war broke out. The Hungarians wanted the forces there reduced but they couldn't break ranks with the Soviets. Anyhow, the problem was that, at the outset, NATO and the Pact disagreed over a division of the participants into countries which would reduce their forces and those which were so-called flanks, like Turkey and Norway on our side and Bulgaria on the Soviet side. The flanks would not be called on to reduce their forces.

The talks started off with a controversy about the status of Hungary which went on for six months and because there was no agreement on what category the delegates should fit into, preventing delegates from having a single plenary session of these negotiations. The Belgians, who were very strong for progress in these talks, insisted there should be plenary sessions, but there was insistence that there couldn't be any plenary sessions until the status of these countries was established and whether they were going to reduce their forces or not reduce their forces.

The agenda was a conservative one: No concessions, no mistakes in these talks; maintain Western unity and preferably not have any results out of these negotiations unless the Pact wants to surrender. At the end of six months, Dean was told that before the preparatory talks began, Kissinger had made a secret deal with Brezhnev to accept that Hungary would not be a direct participant and the Hungarian forces and the Soviet forces in Hungary would not be reduced in these negotiations.

The basic purpose of the MBFR negotiations was to reduce the very large numerical superiority of the Warsaw Pact forces to a position of equality slightly below the level of the NATO forces. The proposal was to lead off with quite a minor reduction of American forces of about 20,000, in return for a reduction of many more Warsaw Pact forces. This goal was pursued for over 16 years. Dean was with it for 8 or 9 years.

When the successor Conventional Forces in Europe (CFE) talks were held, Shevardnadze as the Russian foreign minister stated that the purpose of the talks should be equality at a level lower than the level of the weaker side. He accepted the MBFR objective. Since the Soviets then for the first time exchanged data on their forces showing huge Warsaw Pact numerical superiority in the first session of that negotiation, he conceded the point that the MBFR had been vainly attempting to get for 16 years.

Dean was first in charge of the interagency talks here in the State Department, then of the preparatory talks in Vienna, and then the negotiations proper took place. Stanley Resor was designated as the head of the delegation and Dean was his deputy and remained in that position until 1978, when he became head of delegation.

There was a very tight and apprehensive view of negotiating this kind of subject matter with the Soviets and the Warsaw Pact. The National Security Council, under Kissinger and a series of people that ran the actual coordination, kept a very firm grip on the preparations. Dean was in favor of building down the military confrontation in Europe and believed the time had come to make a real try for this with the Soviets to parallel our relative success in building down the political confrontation in the Berlin negotiations. In the minds of the administration, the talks were designed to fend off Senator Mansfield's push for withdrawal of U.S. forces from Europe and it did not have any serious interest in moving ahead on this front, certainly not before they finished the SALT talks, which were Kissinger's main interest. For example, he didn't want the Germans running out in front on this conventional reduction before he completed agreements with the Soviets. He didn't want to advance the so-called option three, the proposal for reduction of tactical nuclear aircraft to deliver nuclear bombs, until he had the SALT agreement locked in. He always had a kind of hold back attitude toward MBFR: Maintain allied unity, keep the talks going but don't do anything in them.

From the Soviet point of view, it wasn't until 17 years after the start of the MBFR talks that Gorbachev agreed to figures and inspections. In the first day of the new Conventional Forces in Europe (CFE) negotiations the Soviets laid down the data which showed that they did have numerically a lot more than we did, and they accepted the principle which had been vainly pushed for 17 years, that there would be reductions to an equal level below that of the weaker side and also verification. They gave away the entire game that had been pushed for in the MBFR talks in the beginning of CFE.

1.11 Post-Foreign Service

Dean returned to Washington and finally ended up at the Carnegie Endowment for International Peace, retiring from the Foreign Service in 1982. In 1984 he accepted an offer from the Union of Concerned Scientists, analyzing European security developments and related issues. First the missile crisis, the Pershings and the SS-20s, and Germany. Speaking from experience, Dean wrote about the new CFE talks. He followed Gorbachev's course and went to Russia several times for conferences and found that there was life outside the Foreign Service.

Willow Spring Farm, the Dean Family home in Great Falls, Virginia, June 2010. *Source and copyright* Private photo collection of the Dean family

The initial strong reading was that Gorbachev was a fake. That was the predominant administration view, or at least Washington view. Dean thought that a man who would actually unilaterally pull back sizable forces from East Germany was no fake and at least had some claim to be taken seriously. He went to the first Geneva meeting between Reagan and Gorbachev as a representative of a nongovernmental organization (NGO), with Admiral Lee, a longtime nuclear disarmer. Gorbachev talked himself out of the job from the point of view of the Russian establishment. His return from Foros was televised after he had been under the supervision of the coup people for 3 or 4 days. He showed he still was a convinced communist. He spoke of communism with a human face and that it would still some day prevail. Yeltsin was giving him an awful hard time in these sessions.

Jonathan and Theodora Dean at their home in Great Falls, Virginia, June 1985. *Source and copyright* Private photo collection of the Dean family

At the Union of Concerned Scientists Dean worked principally on disarmament. He worked on proposals for what later became the CFE talks and developed quite a few proposals on nuclear disarmament. The NGOs have three main areas of activity, at least the ones in the arms control field. Working with the administration, giving them ideas and concepts. Many of them have already thought of these ideas, but it is often more convenient to pass it around coming from an NGO, than to stick out their own neck for them. They also work with Congress. On most controversial public issues, the discussion is between NGO's, Congress and the administration.

Sometimes, the State Department picks up the ideas of NGOs and presents them as their own. They circulate within the administration some paper or proposal which they might themselves believe is a good idea but announce produced by an NGO. Often, the NGOs are ahead of the curve or of the actual pace of developments and try to look ahead and produce suggestions that will open the way to rational and effective policy.

1.11.1 Global Action to Prevent War

In 1999, together with Saul Mendlovitz and Randall Forsberg, Jonathan Dean founded *Global Action to Prevent War* (GAPW). It was developed as a UN-based, transnational network of civil society, academic and diplomatic partners dedicated to practical measures for reducing levels of global conflict and to removing institutional and ideological impediments to addressing armed violence, mass atrocities and severe human rights violations at the earliest possible stages.

All of Jonathan Dean's life experiences went into his belief in GAPW, and his firm conviction that that armed conflict could be prevented. His belief in the importance of the United Nations continued through his life.

In 2007 Jonathan Dean retired from the Union of Concerned Scientists. He continued to write and to work with GAPW. On June 15, 2014 he would have celebrated his 90th birthday. He passed away five months earlier on 14 January 2014 in Meza, Arizona.

Amb. Jonathan Dean with his daughter Theodora Schiro, in Mesa, Arizona, December, 2012. *Source and copyright* Private photo collection of the Dean family (Tds 3.23.14)

Chapter 2
Bibliography of Jonathan Dean's Work

2.1 Books

Ending Europe's Wars—The Continuing Search for Peace and Security, New York: The Twentieth Century Fund Press, 1994.

Meeting Gorbachev's Challenge—How to Build Down the NATO-Warsaw Pact Confrontation, New York: St. Martin's Press, 1989.

Watershed in Europe—Dismantling the East–West Military Confrontation, Lexington, Mass.: D.C. Heath and Company, Lexington Books, October 1986.

2.2 Co-Authored Books and Reports

The Nuclear Turning Point—A Blueprint for Deep Cuts and De-Alerting on Nuclear Weapons, co-author in Harold A. Feiveson, ed., Washington, D.C.: Brookings Institution Press, 1999.

Options and Opportunities: Arms Control and Disarmament for the 21st Century, with Jeffrey Laurenti, New York: The United Nations Association of the United States of America, Inc., 1997.

The Road Beyond Start: How Far Should We Go, Washington, D.C.: The Atlantic Council of the United States, 1997.

The Final Stage of Nuclear Control, College Park, MD: Center for International and Security Studies at Maryland, PRAC Paper No. 10, August 1994.

Nuclear Security in a Transformed World with Kurt Gottfried, Cambridge, MA: Union of Concerned Scientists, September 1991.

H. G. Brauch and T. Grimwood (eds.), *Jonathan Dean*,
SpringerBriefs on Pioneers in Science and Practice 19,
DOI: 10.1007/978-3-319-06662-2_2, © The Author(s) 2014

German Unity and the Future European Order of Peace and Security, with Horst Afheldt, Malcolm Chalmers, Jonathan Dean, Hans Günter Brauch: AFES-PRESS Report No. 38, Mosbach: AFES-PRESS, 1990.

The INF Treaty and the Future of Western Security with Peter Clausen, Cambridge, Mass.: Union of Concerned Scientists, January 1988.

2.3 Chapters and Articles

"Rethinking Security in the New Century—Return to the Grotean Pattern" (will be published in 2014 in Chinese), in: Brauch, Hans Günter; Oswald Spring, Úrsula; Mesjasz, Czeslaw; Grin, John; Cheng. Liu (Eds.): *Globalization and Environmental Challenges: Reconceptualizing Security in the twenty first century* (Nanjing: Nanjing Press Company, 2014).

"Rethinking Security in the New Century—Return to the Grotean Pattern", in: Brauch, Hans Günter; Oswald Spring, Úrsula; Mesjasz, Czeslaw; Grin, John; Dunay, Pal; Behera, Navnita Chadha; Chourou, Béchir; Kameri-Mbote, Patricia, Liotta, P.H. (Eds.): *Globalization and Environmental Challenges: Reconceptualizing Security in the twenty first century.* Hexagon Series on Human and Environmental Security and Peace, vol. 3 (Berlin–Heidelberg–New York: Springer-Verlag, 2008): 3–6.

"Finding the Missing Element in UN Peacekeeping," in *University of Missouri—Peace Studies Review*, Vol. 2, No. 1&2, Spring and Fall 2006, 3–8.

"International Threats to Competitiveness," in *International Journal of Commerce and Management*, Vol. 13, No. 3&4, 2003, 1–5.

"Ending Wars," in Majid Tehranian and David W. Chappell, eds., *Dialogue of Civilizations: A New Peace Agenda for a New Millennium*, London and New York: I.B. Tauris, 2002, 179–190.

"Security," in Diana Ayton-Shenker ed. *A Global Agenda: Issues Before the 57th Assembly of the United Nations*, Lanham, New York, Boulder, and Oxford: Rowman & Littlefield Publishers, Inc., 2002–2003 Edition, 101–122.

"Arms Control and Disarmament," in Diana Ayton-Shenker and John Tessitore eds., *A Global Agenda—Issues Before the 56th Assembly of the United Nations*, Lanham, New York, Boulder, and Oxford: Rowman & Littlefield Publishers, Inc., 2001–2002 Edition, 61–80.

"Missile Defense a Threat," with Jonathan Granoff in *Chicago Sun-Times* (opinion), August 2001.

"OSCE and NATO: Complementary or Competitive Security Providers for Europe," IFSH in collaboration with Pal Dunay, Adam Rotfeld, and Andrei Zagorski, eds., *OSCE Yearbook 1999*, Vol. 5, 1999, 429–434.

"Global Action to Prevent War: A Programme for Government and Grassroots Efforts to Stop War, Genocide and Other Forms of Deadly Conflict," with Randall Forsberg and Saul Mendlovitz in *Medicine, Conflict, and Survival*, London: Frank Cass & Co. Ltd., Vol. 16, November 1999, 108–116.

"Star Wars Return," in *Sojourners*, July–August 1999, 12.

"Global Action to Prevent War," with Randall Forsberg and Saul Mendlovitz in *Boston Review*, Vol. 24, No. 1, March 1999, 4–10.

"NATO Enlargement: Coping with Act II," in Ted Galen Carpenter and Barbara Conry, eds., *NATO Enlargement—Illusions and Reality*, Washington, D.C.: CATO Institute, 1998, 121–127.

"Expansion Now is Unnecessary," in Kenneth Thompson, ed., *NATO Expansion,* Lanham, MD: University Press of America, Inc., 1998, 109–112.

"Completing a Regime for Deep Cuts in Nuclear Forces," in *Brassey's Defence Yearbook*, London: Brassey's 1998, 267–294.

"Step-by-Step Control over Ballistic and Cruise Missiles," in *Disarmament Diplomacy*, London: The Acronym Institute, Issue Number 31, October 1998, 3–11.

"De-alerting: A Move Towards Disarmament," in *UNIDIR Newsletter*, Geneva: United Nations Institute for Disarmament Research, Number 38, August 1998, 9–13.

"The Case Against NATO Expansion," in *Current History-A Journal of Contemporary World Affairs*, Vol. 97, No. 617, May 1998, 132–136.

"Why not include everyone in NATO," in *Center for War, Peace, and the News Media*, Boston, Mass.: Global Reporting Network, March 1998, 35–37.

"Global Action To Stop 'Small' Wars," in *Disarmament Diplomacy*, London: The Acronym Institute, Issue Number 22, January 1998, 2–7.

"The NATO Mistake," in *The Washington Monthly*, Vol. 29, No. 7, July–August 1997, 35–37.

Comments on the "NATO-Ukraine Charter," in *Berlin Information-center for Transatlantic Security (BITS)* Briefing Note 97.1, July 1997.

"A Stronger U.N. Strengthens America," in George A. Lopez and Nancy Myers, eds., *Peace and Security*, Lanham, MD and Oxford: Rowman & Littlefield Publishers, Inc., 1997, 173–182.

"The USA and the OSCE: Still a Morganatic Union," in IFSH in collaboration with Pal Dunay, Adam Rotfeld, and Andrei Zagorski, eds., *OSCE Yearbook 1997,* Vol. 3. 1997, 39–43.

"No NATO expansion now," in *The Bulletin of the Atomic Scientists*, Vol. 52, No. 3, May–June 1996, 18–19.

"The CFE Treaty Under Attack," *in Disarmament Times*, New York: NGO Committee on Disarmament, Inc., Vol. XIX, No. 2, April 1996.

"Deterrence and Diplomacy," in *Ballistic Missile Defense in Perspective*, Washington, D.C.: The Coalition to Reduce Nuclear Danger, April 1996, 24–27.

"Taking Down the Guns," in *War Report-bulletin of the institute of war and peace*, No. 39, February–March 1996.

"The Year's Progress in Arms Control and Disarmament," in J. Marshall Beier and Steven Mataija, eds., *Verification, Compliance and Confidence Building: The Global and Regional Interface*, Toronto, Canada: Centre for International and Security Studies, 1996, 1–19.

"The OSCE 'Code of Conduct on Politico-Military Aspects of Security': A Good Idea, Imperfectly Executed, Weakly Followed-up," in the IFSH in collaboration with Pal Dunay, Adam Rotfeld, and Andrei Zagorski, eds., *OSCE Yearbook 1995/1996*, Vol. ½, 1995/1996, 291–298.

"The Lessons of War," in *Foreign Service Journal*, Vol. 72, No. 12, December 1995, 32–39.

"Rules Needed Before Deploying First US Defense," in *Foreign Service Journal*, Vol. 72, No. 12, December 1995, 32–39.

"La Sicurezza Globale," with Randall Forsberg in *Technology Review*, N. 84–85, Anno 8, Ottobre–Novembre 1995, 45–47.

"The Road to Zero," with Randall Forsberg in *Technology Review*, Vol. 98, No. 6, August–September 1995, 74.

"Building a Cooperative Security System," in *Disarmament Times,* New York: NGO Committee on Disarmament, Inc., August 1995, 6.

"Deterrence and the Nuclear Arms Race," in *Disarmament Times,* New York: NGO Committee on Disarmament, Inc., August 1995, 3.

"East Europe Membership Would Weaken Security Alliance," in *Foreign Service Journal*, Vol. 72, No., August 1995, 22–25.

"Losing Russia or Keeping NATO: Must We Choose," in *Arms Control Today*, Washington, D.C.: Arms Control Association, Vol. 25, No. 5, June 1995, 3–7.

"A Stronger U.N. Strengthens America," in *The Bulletin of the Atomic Scientists*, Vol. 51, No. 2, March–April 1995, 45–54.

"Unter welchen Bedingungen kann eine Europäische Sicherheitsgemeinschaft errichtet werden" in *Unsere Gemeinsame Zukunft*, Band 88, Aufl. 1994/95, 229–234.

"Wege zur gesamteuropäischen Sicherheit," in *S+F Vierteljahresschrift fur Sicherheit und Friede,* Jahrgang 13, Heft 4, 4/1995, 216–217.

"Neutralisierung von Atomwaffen: Ein Vorschlag zur Reduzierung der politischen Bedeutung von Kernwaffen," in *S+F Vierteljahresschrift für Sicherheit und Frieden*, Jahrgang 13, Heft 3, 3/1995, 170–175.

"The Final Stage of Nuclear Arms Control," in *The Washington Quarterly*, Vol. 17, No. 4, Autumn1994, 31–52.

"Yugoslav Errors, Peacekeeping and the American National Interest," in *The Aspen Institute Quarterly*, Vol. 6, No. 4, Autumn 1994.

"Arms Control and Nonproliferation Issues," Kenneth W. Thompson, ed., in *Presidents and Arms Control*, Lanham, New York, and London: University Press of America, University of Virginia, The Miller Center (co-publisher), 1994, 189–212.

"Using Arms Control to Promote NOD in Europe," in Bjorn Moller and Hakan Wiberg, eds., *Non-Offensive Defence for the Twenty-First Century*, Boulder, San Francisco, & Oxford: Westview Press, 1994, 37–47.

"Keep Cutting Nuclear Weapons," in *The Christian Science Monitor*, August 31, 1994, 19.

"The Final Stage of Nuclear Arms Control—and How to Achieve It," in *Disarmament—A periodic review by the United Nations*, Vol. XVII, No. 2, 1994, 94–107.

"Unter welchen Bedingungen kann eine Europäische Sicherheitsgemeinschaft errichtet werden," in *S+F Vierteljahresschrift für Sicherheit und Frieden*, Jahrgang 11, Heft 3, 1993, 126–128.

"Comprehensive Control over Nuclear Weapons," edited by Ivo Daalder and Terry Terriff in *Arms Control: Contemporary Security Policy*, London: Frank Cass & Co. Ltd., Vol. 14, No. 1, April 1993, 240–251.

"Comprehensive Control over Nuclear Weapons," in Ivo H. Daalder and Terry Terriff, eds., *Rethinking the Unthinkable—New Directions for Nuclear Arms Control*, London: Frank Cass & Co. Ltd., 1993, 240–251.

"Expanding the Security Council Role in Blocking the Spread of Nuclear Weapons," in *Transnational Law & Contemporary Problems*, Vol. 2, No. 2, Fall 1992, 587–603.

"CFE and Beyond," with Randall Watson Forsberg in *International Security*, Vol. 17, No. 1, Summer 1992, 76–120.

"A Program for World Nuclear Security," with Kurt Gottfried in *Union of Concerned Scientists*, Cambridge, Mass.: February 1992.

"Components of a Post-Cold War Security System for Europe," in Hans Günter Brauch and Robert Kennedy, eds., *Alternative Conventional Defense Postures in the European Theater*, Vol. 2, 1992, 45–68.

"Constraining technological weapons innovation in the post-Cold War environment", in Hans Günter Brauch, H. van der Graaf, John Grin and Wim Smit, eds., *Controlling the Development and Spread of Military Technology*, Amsterdam: VU University Press, 1992, 35–41.

"Requirements for successful negotiations on force reductions in Europe," in *East–West arms control: Challenges for the Western Alliance*, London: Routledge, 1992, 49–59.

"The Post-Cold War Security System in Europe—An Evaluation, in *Cornell International Law Journal*, Vol. 24, No. 3, 1991, 457–468.

"Nuclear Security in a Transformed World," with Kurt Gottfried in *Arms Control Today*, Vol. 21, No. 9, November 1991

"Arms Reduction Blueprint," edited by Hans Gunter Brauch and J.W. Goethe Fankfurt in *AFES–Press Report No. 45*, Germany: Session 1.4, October 1991, 75–81.

"Coalitions for Regional Crises," in *The Bulletin of the Atomic Scientists*, Vol. 47, No. 5, June 1991, 33.

"Arms Control on the Korean Peninsula: How Is the European Experience Applicable," in *The Korean Journal of Defense Analysis*, Vol. III, No. I, Summer 1991, 67–84.

"Conventional and Nuclear Forces in Europe: the Importance of the Vienna Talks," in *Reducing Nuclear Arsenals*, Hampshire and London: MacMillan Academic and Professional LTD, 1991, 225–240.

"Europe's New Security Blanket," in *Pacific Research*, Vol. 4, No. 1, February 1991, 7–10.

"Arms Control on the Korean Peninsula: How is the European Experience Applicable," in *The Korean Journal of Defense Analysis*, Issue 3, January 1991, 67–84.

"The End of Verification," in Heather Chestnutt and Michael Slack, eds., *Verifying Conventional Force Reductions in Europe: CFE I and Beyond*, Toronto, Canada: Centre for International and Strategic Studies, York University, 1991, xv–xxii.

"Defining Long-term Western Objectives in CFE," in *The Washington Quarterly*, Vol. 13, No. 1, Winter 1990, 169–182.

"Conventional Arms Reduction in Europe: Past, Present, and Future, " in *Arms Control in Korean Peninsula—What Lessons can we learn European experiences*, Seoul, Korea, Institute of Foreign Affairs and National Security, December 1990, 23–36.

"The Emerging Framework for European Security," in *Reforging European Security: From Confrontation to Cooperation*, Boulder, CO: Westview Press, December 1990, 85–101.

"CFE: A Good First Step," in *Arms Control Today*, Vol. 20, No. 10, December 1990, 3.

"Peace and Security in 2050: What we need to do NOW," edited by Sophia Peterson and Virgil A. Peterson in *FACDIS*, Morgantown, WV: The West Virginia Consortium for FACDIS, November 1990, 51–57.

"The CFE negotiations, present and future," in *Survival*, Vol. XXXII, No. 4, July–August 1990, 313–324.

"Building a Post-Cold War European Security System," in *Arms Control Today*, Vol. 20, No. 5, June 1990, 8–12.

"The CFE Negotiations", in *SNU*, Nyborg, Denmark, The Danish Commission on Security and Disarmament, April 1990, 21–28.

"Negotiated Force Cuts in Europe: Overtaken by Events," in *Arms Control Today*, Vol. 19, No. 10, January 1990, 12–16.

"Nicht-offensive Verteidigung bei den laufenden Wiener Abrüstungsverhandlungen," in *Die Zukunft des Friedens in Europa*, Carl Hanser Verlag, January 1990, 293–299.

"Alternative Defence—Answer to the Bundeswehr's Post-INF Problems," in Stephen F. Szabo, ed., *The Bundeswehr and Western Security*, Houndsmill, Basingstoke, Hampshire, and London: MacMillan Academic and Professional LTD, 1990, 188–216.

"Beyond the INF Treaty," Kenneth W. Thompson, ed., in *Arms Control: Moral, Political and Historical Lessons*, Lanham and London: University Press of America, 1990, 29–51.

"Immer noch vonnöten: ein gemeinsames Zentrum zur Risikominderung von NATO und Warschauer Pakt," in *Abrüstung und Konventionelle Stabilität in Europa*, Germany: Nomos Verlagsgesellschaft: Baden-Baden, Band 63, 1990, 135–146.

"Is NATO's Past Also Its Future," in *The Declining Hegemon: The United States and European Defense*, Westport, CT: Greenwood Press, 1990, 232.

"The Vienna Force Reduction Talks: Moving Towards Deep Cuts," in *The Foundations of Defensive Defence*, Houndsmill, Basingstoke, Hampshire, and London: MacMillan Academic and Professional Ltd., 1990, 195–203.

"Berlin gestern, heute und morgen," in *Berlin: Vom Brennpunkt der Teilung zur Brücke der Einheit*, Bundeszentrale für politische Bildung, Band 288, 1990, 366–381.

"Planning for the Next Round of CFE Talks," in *Defense & Disarmament Alternatives*, Vol. 2, No. 11, November 1989, 1–2.

"Conventional Talks: A Good First Round," in *The Bulletin of the Atomic Scientists*, Vol 45, No. 8, October 1989, 26–31.

"Can NATO Agree on Arms Control," *in Technology Review*, Cambridge, Mass.: The Massachusetts Institute of Technology, October 1989, 58–68.

"Cutting Back the NATO-Warsaw Treaty Confrontation," in *Disarmament*, Vol. XII, No. 2, Summer 1989, 18–25.

"How to Reduce NATO and Warsaw Pact forces," in *Survival: Global Politics and Strategy*, Vol. 31, Issue 2, March–April 1989, 109–122.

"Towards a Cooperative Security Regime in Europe," in Kurt Gottfried, ed., Ithaca, NY: Cornell University Peace Studies Program, August 1989.

"Alternative Defense—Answer to the Bundeswehr's Post—INF Problems," in Steven Szabo, ed., Federal German Bundeswehr, June 1988.

"INF- Vertrag und europäische Sicherheit," with Peter Clausen in *Perspektiven für Sicherheit und Zusammenarbeit in Europa*, Bundeszentrale für politische Bildung, Schriftenreihe Band 268, 1988, 228–238.

"Will Negotiated Force Reductions Build Down the NATO—Warsaw Pact Confrontation," in *The Washington Quarterly*, Vol. 11, Issue 2, Winter 1988, 69–84.

"Can NATO Unite to Reduce Forces in Europe," in *Arms Control Today*, Vol. 18, No. 8, October 1988, 11–18.

"Will the Two German States Solve the Problem of European Security," in *SAIS Review*, Vol. 8, No. 2, Summer-Fall 1988, 173–190.

"The New NATO-WTO Force Reduction Talks: An Optimal Outcome," in *Defense & Disarmament Alternatives*, Vol. 1, No. 5, July–August 1988, 1–4.

"Conventional Arms Talks: now is the time," in *Journal of European Nuclear Disarmament*, Issue 33, May–June 1988, 4–6.

"Berlin in a Divided Germany: An Evolving International Regime," in Alexander George, Philip Farley, and Alexander Dallin, eds., *U.S.—Soviet Security Cooperation—Achievements, Failures, Lessons*, New York and Oxford: Oxford University Press, 1988, 83–105.

"How Can Arms Control Raise the Nuclear Threshold in Europe," in P. Terrance Hopmann and Frank Barnaby, eds., *Rethinking the Nuclear Weapons dilemma in Europe*, Houndsmill, Basingstoke, Hampshire, and London: MacMillan Academic and Professional Ltd., 1988, 291–301.

"The New NATO-PACT Force Reduction Talks—An Optimal Outcome," in *Union of Concerned Scientists*, Washington, D.C.: May 1988, 1–40.

"The NATO-WARSAW PACT Confrontation in the Twenty-First Century: Rough Model for an Optimal Force Posture," Union of Concerned Scientists, Washington, D.C.: May 1988, 1–38.

"Twelve Steps for Fundamentally Improving America's External Environment," Union of Concerned Scientists, Washington, D.C.: March 1988, 1–11.

"The Exclusionary Zone: An American Approach," in *The New Force Reduction Negotiations in Europe: Problems and Prospects*, Washington, D.C.: American Association for the Advancement of Science, February 1988, 5–8.

"Alternative defence: answer to NATO's Central Front Problems," in *International Affairs*, Vol. 64, No. 1, Winter 1987/8, 61–82.

"Military Security in Europe," in *Foreign Affairs*, New York, NY: Council of Foreign Relations, Inc., Vol. 66, No. 1, Fall 1987, 22–40.

"Chemical Weapons in Europe," in Brad Roberts, ed., *Chemical Warfare Policy: Beyond the Binary Production Decision*, Washington, D.C.: The Center for Strategic and International Studies—Georgetown University, Vol. IX, No. 3, 1987, 11–17.

"Turning Point in Europe," in *Foreign Affairs*, New York, NY: Council of Foreign Relations, Inc., August 1987.

"The INF Agreement: Pluses and Minuses for Western Security," in *Arms Control Today*, Washington, D.C.: The Arms Control Association, July–August 1987

"Gorbachev's arms control moves," in *Bulletin of the Atomic Scientists*, Vol. 43, No. 5, June 1987, 34–40.

"Berlin: Enduring Stability or Shaky Interim," in *The Washington Quarterly*, Vol. 10, No. 3, Summer 1987, 87–101.

"US Approaches to Arms Control: Contentious, Erratic, and Inadequate," in *Harvard International Review*, Vol. IX, No. 5, May–June 1987, 14–16.

"L'Alliance peut-elle survivre a sa reussite (relative)," in *Revue de l'OTAN*, Vol. 34, No. 6, December 1986, 14–21.

"Chemical Weapons in Europe," in *Arms Control Today*, Vol. 16, No. 6, September 1986

"Assessing the Warsaw Pact Threat to NATO's Central Front," in *The International Spectator*, Vol. XXI, No. 3, September 1986, 26–34.

"Ein Krisenmanagementzentrum in Europa—ein konkreter Vorschlag einer Vertrauens—und Sicherheitsbildenden Maßnahme für Wien und Stockholm," in Hans Günter Brauch, ed. *Vertrauensbildende Maßnahmen und Europäische Abrüstungskonferenz*,, Militärpolitik and Rüstungsbegrenzung, Vol. 9. Gerlingen. Bleicher Verlag, Band 9, 1986, 329–338.

"Erhöhung der nuklearen Schwelle in Europa—Ein konkreter Vorschlag," in *Die Neue Gesellschaft/Frankfurter Hefte*, Nr. 2, February 1986, 128–132.

"A Five-Year Program for The United States: The Most Urgent Steps," in *East–West Arms Control*," in *Union of Concerned Scientists*, Cambridge Mass.: 1985.

"Superpower Survey," in *Foreign Service Journal*, December 1985, 9–14.

"What They Could, Should Have Done About Arms," in *Los Angeles Times*, Part II, November 1985, 5.

"Directions in Inner-German Relations," in *Orbis*, Fall 1985, 609–632.

"East–West conflict management through European arms control negotiation," in *International Journal XL*, Autumn 1985, 714–735.

"Will NATO Survive Ballistic Missile Defense," in *Journal of International Affairs*, Vol. 39, No. 1, Summer 1985, 95–105.

"A Five-Year Program for the United States: The Most Urgent Steps in East-West Arms Control," in *Toward A New Security: Lessons of the Forty Years Since Trinity*, Cambridge, Mass.: Union of Concerned Scientists, July 1985, 55–58.

"MBFR: Past and Future," Derek Paul, ed., *Defending Europe: in Options for Security*, London and Philadelphia: Taylor & Francis Ltd, 1985, 82–91.

"Verifying Force Reductions and Confidence-Building Measures, in *Symposium on Verification of Disarmament in Europe—Stockholm 1985*, Swedish National Defence Research Institute, 1985, 101–125.

"Arms Control Prospects and Problems: The Next Five Years," in John O'Manique, ed., *A Proxy For Trust*, Ottawa, Canada: The Norman Paterson School of International Affairs, Carleton University, 1985, 1–12.

"Break the Deadlock: A Way To End the U.S.—Soviet Stalemate," in *The Sun*, October 26, 1984.

"Conventional Retaliation into Eastern Europe," with Samuel Huntington, B. Bruce-Briggs, and Jeffrey Record in *International Security*, Vol. 9, No. 1, Summer 1984, 203–217.

"Debating the Role of Nonstrategic Nuclear Weapons," in *Environment*, Vol. 26, No. 6 July–August 1984, 7–13 (33–38).

"How to Lose Germany," in *Foreign Policy*, No. 55, Summer 1984, 54–72.

"An East–West Arms Control Agreement This Year," in *Carnegie Endowment for International Peace*, Washington, D.C.: February 23, 1984, 1–4.

"Federal Germany after the Euromissiles," in *The Bulletin of the Atomic Scientists*, Vol. 39, Issue 10, December 1983, 31–38.

"Battlefield Nuclear Weapons: Defusing the NATO-Warsaw Pact Confrontation in Central Europe," in *Arms Control Today*, Vol. 13, No. 8, September 1983, 1, 6–8.

"MBFR: From Apathy to Accord," in *International Security*, Vol. 7, No. 4, Spring 1983, 116–139.

"Negotiation by Increment," in *Foreign Service Journal*, July–August 1983, 26–34.

"Uber den Ersteinsatz hinaus," in *Die Neue Gesellschaft 2*, Jahrgang 30, February 1983, 74–108.

"MBFR: Ten Years of Negotiating Security in Europe," in *Arms Control Today*, Vol. 12, No. 8, September 1982, 1, 8.

"Beyond First Use," in *Foreign Policy*, No. 48, Fall 1982, 37–53.

Jonathan Dean on holiday with his wife Theodora (1950, Limburg, Germany). *Source* Private photo collection of the Dean family

Part II
Texts by Jonathan Dean on Detente and Arms Control in Europe in the 1980s

Ambassador Jonathan Dean (*center*) in a conversation with Prof. Dr. Carl Friedrich von Weizsäcker (to the *left*) and Ambassador Roth, German Foreign Office (*right*) during a conference organized by the Social Democratic Party of Germany on disarmament issues in 1977. *Source* Personal photo collection of Ms. Theodora Schiro, daughter of Jonathan Dean. Reprinted with permission

Chapter 3
Beyond First Use (1982)

In recent months several prominent observers have suggested that the United States and its *North Atlantic Treaty Organization* (NATO) allies consider relinquishing their strategy of possible first use of tactical or intermediate-range nuclear weapons to counter a Warsaw Pact conventional attack that threatened to overwhelm Western Europe.[1] Those raising this issue include West German Christian Democratic leader Kurt Biedenkopf, several respected British authorities on defense, conservative American commentator Irving Kristol, and four former U.S. officials of unusual eminence—McGeorge Bundy, George Kennan, Robert McNamara, and Gerard Smith.

The majority response, official and otherwise, has been unambiguously negative. In April 1982 then Secretary of State Alexander Haig. Jr. argued that abandoning the first use strategy would mean reintroducing the draft, tripling the size of the U.S. armed forces, and putting the economy on a wartime footing. In May 1982 the NATO foreign ministers officially re-endorsed the first use strategy.

The main arguments advanced for retention of the NATO first use strategy are as follows:

- The first use strategy still has military value; it may represent a decisive obstacle to a Soviet attack on Western Europe.
- Abandoning the strategy would sever the link between defense of Western Europe and possible use by the United States of strategic nuclear weapons against the Soviet Union. This linkage is considered the ultimate deterrent to Soviet attack against Western Europe.
- Warsaw Pact superiority over NATO conventional forces is so great that achievement by NATO of a conventional balance with Pact forces would cost

[1] This text was first published as: "Beyond First Use", in: *Foreign Policy*, No. 48 (Fall 1982): 37–53. The permission to republish this text was granted by Mr. Curry of Foreign Policy Magazine on 4 April 2014.

H. G. Brauch and T. Grimwood (eds.), *Jonathan Dean*,
SpringerBriefs on Pioneers in Science and Practice 19,
DOI: 10.1007/978-3-319-06662-2_3, © The Author(s) 2014

too much. Moreover, in the event of actual conflict, conventional war would be as destructive of human life as nuclear war.
- Hence, continued reliance on nuclear deterrence, of which the first use strategy is an essential part, remains the best guarantor of peace in Europe.

Because most NATO governments strongly hold these views, NATO will retain the first use strategy for the next few years at least. Yet the issue does not end with official confirmation of the strategy. Proposals to drop the first use strategy have come on the heels of intense and continuing controversy in Western Europe over deployment of U.S. intermediate-range nuclear missiles. Both the deployment of new U.S. missiles and the first use strategy that these missiles might someday carry out are themselves specific applications of the concept of nuclear deterrence to West European defense. Thus the real issue underlying the current debate is the role of nuclear deterrence in the defense of Western Europe and the more general question of what is the best defense strategy for Western Europe.

This West European debate over NATO strategy is coming into full swing and will not come to rest until a new and broader consensus is reached on how best to defend the security of Western Europe. Examination of the rationale for the first use strategy advanced by its proponents can provide an opportunity to evaluate NATO's current defense strategy. The key question is whether the degree of emphasis it now places on nuclear deterrence can receive adequate support over the long run from West European public and parliamentary opinion and if not whether it is possible to reach some tentative conclusions about the feasibility of relinquishing or of modifying that strategy.

3.1 Unconvincing Strategy

The military value of the first use strategy can be analyzed in terms of three criteria: its value in deterring the Soviet Union from attack; its value in providing general guidance for the actual conduct of defense in the event of conflict and thus for organizing and equipping NATO armed forces during peacetime; and its value in reassuring the population being defended that the strategy provides a reasonable chance of preventing war and of protecting that population if war comes.

The deterrent effect of NATO's first use strategy on Soviet leaders cannot lie reliably measured. No one in the West knows whether Soviet leaders have at any point seriously contemplated military attack against Western Europe. The prudent assumption is that since the Soviets have never attacked Western Europe, the first use strategy has had some deterrent effect and will continue to have this effect. At the same time, it appears logical to assume that deterrence has lost some of its former weight as the strategic balance has shifted from dear U.S. superiority to rough equality between the superpowers. It is possible that Soviet leaders may now believe that in a military crisis involving only Europe, the President of the United States would seek to protect the American heartland by confining the struggle to Europe.

Thus he would decide in the end not to use nuclear weapons at all since any use could easily escalate to a strategic exchange of nuclear weapons between the United States and the USSR.

It should be noted, however, that deterrence would exist even without the first use strategy. Relinquishing the strategy would not mean withdrawal of American nuclear armaments from Western Europe. And, in the event of an overwhelming Warsaw Pact conventional attack, Soviet leaders would remain uncertain about how the West would respond. In addition, the prospect of engaging in a global conventional war with the United States and its allies is itself a very powerful deterrent to Soviet attack on Western Europe. Such a conflict would raise the possibility that Moscow might lose control over Eastern Europe, that the Chinese might intervene, and even that the Soviet political system itself might suffer.

Regarding the merits of the first use strategy, under the second criterion of guidance for the actual conduct of defense in the event of conflict, one possible first use of American nuclear weapons, recently discussed in the press, could take the form of a signal shot in the air over the Baltic Sea. NATO might take this action in an effort to convince the Warsaw Pact to stop its advance and to enter into political negotiations.

During the period when the United States enjoyed nuclear superiority, a warning shot of this kind would have alerted Soviet leaders to the reality that further advance by Warsaw Pact troops into Western Europe would risk escalation into a nuclear conflict that the West would win because of the U.S. nuclear edge Given rough Soviet-American nuclear equality', however, such an approach has questionable value. A warning shot might still end the fighting; but it is also possible that the Soviets might respond with a signal shot of their own, or continue their advance, or both. A nuclear exchange against ground targets might then take place in Europe, creating disarray among the field armies of both sides. The outcome of further conflict might under those circumstances depend on outside reinforcements from the United States and the Soviet Union, a situation that would accentuate the geographic advantages of the USSR.

The first use strategy falls even further short of meeting the third requirement for a strategy of defense—assuring the population of Western Europe that the chosen strategy provides the best reasonable chance of preventing war and of protecting the population if war comes. The large-scale, highly motivated public demonstrations of the West European peace movement in I VH 1 provided unambiguous evidence that NATO's reliance on nuclear deterrence is no longer convincing to a growing number of those whom it seeks to defend. For adherents of the peace movement, the strategy is beginning to frighten more than it reassures. And according to public opinion polls, wide segments of the general public in NATO's Northern Tier countries—the Federal Republic of Germany, Belgium, the Netherlands. Luxembourg, Denmark, and Norway—share their concerns.

3.2 The End of U.S. Superiority

Years of extensive discussion in the United States of the U.S.–Soviet nuclear balance have brought an inescapable message to the West European public: namely, that U.S. nuclear superiority, on which extended deterrence and the first use strategy have been based, no longer exists. Consequently, large segments of the public, in many cases for the first time, are now looking more closely at what would happen if war broke out in Europe and existing NATO strategy actually had to be applied. They do not like what they see. This reaction is particularly strong in West Germany, where an exposed location and the scaring experience of World War 11 combine to create understandable sensitivity in dealing with the implications of NATO strategy.

The frequently presented evidence that the USSR has achieved parity with the United States in nuclear armaments has increased the number of West Europeans who suspect that in the event of an overwhelming Soviet attack on Western Europe the U.S. president would decide not to use American nuclear weapons against the invading Soviet forces because the Soviets might retaliate with their strategic weapons directly against the United States. At the same time, evidence of Soviet nuclear equality has elicited increasing nervousness in different segments of the West European public for another reason: They fear that if the West was losing in a conventional conflict, the U.S. president would in fact authorize use of American nuclear weapons, but against targets in Eastern Europe rather than in the Soviet Union, and that the Soviets for their own part would respond with nuclear weapons aimed at Western Europe. Thus each superpower would delay resorting to strategic attack on the other's home territory while Europe was devastated by the nuclear exchange.

The NATO decision of December 1979 to deploy modernized American Pershing II missiles and ground-launched cruise missiles in Western Europe was an effort to counter these concerns and to maintain the credibility of the American extended deterrent and of the strategy of flexible response. NATO governments argued that deployment of these missiles would establish a seamless deterrent linking tactical and intermediate-range nuclear armaments with U.S. strategic nuclear weapons. To many West Europeans, however, the logic of the proposed remedy did not appear compelling. They feared that deployment would instead increase the risk of a U.S.–Soviet nuclear exchange in Europe. Rather than reassuring the population of Western Europe, the decision to deploy became a center of controversy, especially in the Northern Tier countries.

Today NATO's nuclear deterrent strategy is no longer credible to a large number of West Europeans whom it seeks to defend and is not acceptable as a means of defense to many of those who do find it credible. Consequently, from the viewpoint of their national interest, Americans must now ask: Does extended deterrence in its present form, which includes the first use strategy with its potential commitment to use American nuclear weapons in the early stages of a conflict in Europe, continue to provide a sound basis for the necessary American contribution to the defense of Western Europe?

After so many years, the strategy of nuclear deterrence is thoroughly identified in West European thinking with the general issue of U.S. support for the defense of Western Europe. Many of the West Europeans who are concerned by the form of American support for West European defense do not wish to reject that support outright. The result is ambivalence in the Northern Tier countries as evidenced by these apparently parallel phenomena: large antinuclear demonstrations with anti-American overtones yet solid majorities in public opinion polls favoring continued alliance with the United States or a public dissatisfaction with NATO nuclear deterrent strategy yet a desire by NATO governments to cling to the first use strategy. In many individuals' frustration and anxiety over the role of American nuclear armaments in the defense of Western Europe and inability to resolve the dilemma it causes have been transformed into resentment against the United States as the donor of this uncertain and potentially annihilating form of protection.

If the current level of abrasion and friction in West European public opinion over NATO's nuclear deterrent strategy continues, the consequences can be serious. Both the willingness of the West European public to sustain large defense budgets and the resolve of West European leaders to resist possible Soviet pressures could decline. This damage to defense morale could in time outweigh whatever gain for the defense of Western Europe deterrence might provide. Already the controversy threatens to contribute to a destabilizing realignment of the West German political parties. And it could ultimately, through a cycle of action and reaction, result in serious pressure for withdrawal of American forces from Western Europe. Conversely, if well handled, the current dissatisfaction in Western Europe over nuclear deterrence could have a productive outcome: It could finally convince the West Europeans, especially the West Germans, to take conventional defense seriously.

For NATO's real and continuing requirement is for a conventional force posture that makes attack by Warsaw Pact forces an irrational course and that, in the event of a Soviet miscalculation, could lead to defeat of the attacking forces. But after establishing their original ambitious conventional force goals in 1952, the NATO allies deliberately revised those goals and decided not to try to match estimated Warsaw Pact strength man for man, tank for tank, aircraft for aircraft. Instead, they placed primary reliance on U.S. superiority in nuclear weapons, with conventional forces playing a lesser role. The reasons for this derision were many, including distrust of excessively large West German forces and the sensitivity of West Germany to the idea of prolonged conventional warfare on its soil. But in large part, the decision reflected the belief of the NATO allies that Warsaw Pact conventional advantages were so great that NATO would find conventional parity too expensive in economic and political terms.

These were NATO's conclusions in the early 1950s. At that time, the Soviet Union did have a marked preponderance in conventional forces because of the postwar disbandment of West European and American forces and the disarmament of defeated Germany. But today such pessimistic conclusions may no longer be valid.

3.3 Comparing the Alliances

Despite the concerns of recent years over trends in Soviet and Warsaw Pact forces, NATO forces in central Europe today have important advantages in specific areas. NATO air forces maintain an impressive edge over Pact air forces in many categories including payload. NATO pilots have logged more flying hours and arc better trained. NATO has about twice as many modern fighter-bombers as the Warsaw Pact. NATO leads in antitank guided missiles and helicopters.

NATO possesses the important advantages of a force defending its own soil: prepared defensive positions, relatively favorable terrain, greater knowledge of its own area, and higher military morale. NATO also benefits from the general requirement that the potential attacker must have considerable overall superiority to insure success. An additional advantage would come from the supportive efforts of the West European civilian population, which in a war would cooperate fully with Western defense forces.

Facing Western forces would be Warsaw Pact troops with a manpower advantage in central Europe of about 175,000 in ground and air force personnel—a level that does not appear to have markedly increased since the 1968 Soviet invasion of Czechoslovakia. Although this margin has a different significance in an arms control context, it does not appear to create a decisive advantage for the Pact in conflict. It is true that Soviet ground forces in the western USSR could be used to reinforce Warsaw Pact troops already in central Europe, but this advantage is offset by the possibility of Western reinforcement by French ground forces in France. Moreover, according to figures published by the Defense Department, since 1972 the United States has increased its ground and air force manpower in West Germany by 38,000 men. It has also carried out a major reorganization of U.S. Army forces in central Europe to increase combat capability by reducing support forces.

NATO military commanders face problems regarding the quality of military' manpower, but so do Soviet commanders. Soviet conscripts in general have a low level of education and mechanical aptitude, which decreases their usefulness in a highly mechanized force. And the pronounced centralism of the Soviet system creates difficulties in training the commanders of small units to show initiative. And some Soviet units in the area arc not equipped with latest model armaments. Over half the Soviet units in central Europe are equipped with the older, less advanced T-62 tank. Western force analysts professionally avoid qualitative estimates. But it is probable that the quality of Soviet ground forces in central Europe, clearly the best in the Warsaw Pact, is considerably lower than that of West Germany's Bundeswehr.

Assessment of the military' balance in central Europe must take into account the actual political situation in Europe today. In the 1950s a large portion of the West European population was loyal to Communist parties subservient to Moscow. This loyalty posed a serious threat to Western defense in the event of conflict, as NATO commanders envisaged large-scale efforts at fifth-column activities. Today the situation is reversed. Following the 1956 Hungarian uprising, the 1968 reform movement in Czechoslovakia, and the 1980–1981 developments in Poland, the locus of political crisis in Europe has shifted from West to East. The Soviet

Union faces a permanent political challenge in Eastern Europe. This challenge raises important questions both about the behavior of civilian populations in Eastern Europe during a conflict and about the military value to the Soviets of the armed forces of Warsaw Pact countries in the area.

The Czechoslovak army remains demoralized and of poor quality. Soviet commanders must have second thoughts about trusting East German conscripts in West Germany for any length of time. The loyalty of the Polish army to Soviet leadership is highly questionable. In the face of these problems, before launching an attack on the West the Soviet Union would probably have to assign a considerable number of its existing forces to insure its lines of communication through Eastern Europe. Militarily, in other words, the Soviet Union is largely on its own in Eastern Europe. Even in the early stages of conflict in Europe, the entire Soviet system in Eastern Europe might be at risk. Awareness of these dangers is itself an important deterrent to any Soviet decision to attack the West.

The most plausible scenario of a Soviet attack on Western Europe—perhaps the one most feared by NATO leaders—is an attack by Soviet forces already in central Europe, with minimum preparation of a week or less and without immediate reinforcement from the Soviet Union. For such an attack, the Soviets would have available 19 divisions in East Germany and perhaps 3–4 divisions in Czechoslovakia, or a total of about 23 divisions. This figure is far short of the 175 Pact divisions so often mentioned in popular discussion of the subject. Twenty-three Soviet divisions constitute a formidable force, but given the advantages of defense, NATO should be able to cope with them. Ixraving out the three French divisions in West Germany, NATO has considerably more manpower in active duty combat units in central Europe than the Soviets have personnel in all their combat units in this area.

Attack by Soviet forces alone after limited preparation is not the only form that conflict in central Europe could take. Some part of non- Soviet Warsaw Pact forces might move forward. Sandwiched between loyal Soviet forces to insure obedience. NATO commanders must take this possibility info account. But even in these circumstances, the value of the East European troops remains questionable. Apart from their uncertain morale and motivation, these troops possess inferior equipment. Nearly all their tanks arc of a design more than 30 years old. They have only a handful of modem third generation fighter-bombers. The actual readiness level of many divisions usually counted among active duty forces is very low and it would take weeks to mobilize these divisions fully. In addition, mobilization of both Soviet and non-Soviet Warsaw Pact forces would be highly visible. Assuming the Western alliance would react accordingly, some 750,000 West German reservists and the reserves of other allies could reinforce active dury Western forces in the area. In short, the West can field a formidable force in central Europe, quite apart from American reinforcements. And the United States has pre-positioned enough equipment for four divisions in West Germany, thus enabling reinforcements to arrive rapidly by air.

This necessarily incomplete evaluation of the East–West military balance in central Europe does not suggest that NATO has no problems in the conventional defense of Western Europe. It does. But from some perspectives, the balance is decidedly less unfavorable for the West than is often assumed.

3.4 A New Defensive Approach

There remain two final objections to relinquishing NATO's no first use doctrine: First, an adequate conventional defense fo. Western Europe would be prohibitively expensive for the West. Second, even if the West were able to shoulder the burden, a conventional defense of Western Europe would result in destruction and loss of life, especially in West Germany, at a level approaching or equaling the carnage of World War II or even of a nuclear war.

Assessing the economic costs of an adequate conventional defense of Western Europe must rest in large part on an evaluation of the military balance in the area. But as argued above, evidence indicates the capabilities of Warsaw Pact forces in central Europe arc not as great as often assumed. Therefore, if NATO countries achieved a 3 % real increase in defense expenditures over a period of 4 or 5 years, the chances of a successful conventional Warsaw Pact attack could be made reassuringly low even under the most improbable, worst case scenarios.

In fact, in recent years NATO's logistics and defense preparations have improved considerably. Important programs for further improvements continue. They include NATO programs to facilitate rapid deployment of U.S. reinforcements by improving the host nations' support capabilities and by enhancing U.S. air transport. NATO is also proceeding with programs of aircraft modernization, with increases in stocks of ammunition and war materiel reserves, antitank guided missiles, and artillery, and with improvements in joint command and control.

All these actions are useful. But even more can be done if NATO is prepared to consider innovative approaches to conventional defense. Several such approaches are now under discussion. One idea was first discussed as the West German forces were being established in the mid-1950s: NATO can make more extensive use of prepared defensive positions in the forward area either with standing forces or with a home guard. This approach is relatively inexpensive. A second improvement calls for wider use of modern technology with precision guided munitions designed to interdict possible Soviet reinforcement. Although more expensive, this approach could be implemented in stages.

An official statement by the government of France that, in the event of Soviet attack, French forces would come to the support of NATO forces would of itself do a great deal to invigorate NATO conventional defense. Such a statement would not require France to rejoin the NATO integrated command. Nor would it cost anything in economic terms. But it would enable NATO political and military leaders, and West European public opinion, to count on the French to help in the event of war rather than merely to hope that they will do so. France would guarantee use of its logistics, airfields, and mobile ground forces to help stop Warsaw Pact attackers after minimum penetration of West Germany. This assurance would make the prospects for successful conventional defense appear more solid to West European public and political opinion.

NATO conventional defense could also be strengthened if a greater number of West Germany's ground force reserves were organized into combat units, creating a reserve division to match each of the present 12 West German divisions on active duty. Six brigades or the equivalent of 2 divisions of such forces already exist: The additional 10 reserve divisions, perhaps 5 armored and 5 light infantry, could have a cadre of 10 % full-time active duty personnel and be equipped in part with used equipment from active duty divisions. If these divisions arc added to West German farces at the rate of 2 a year over a period of 5 years or longer. West Germany could implement this program with a limited increase of its defense budget.

The present concept for conventional defense against Warsaw Pact attack, at least in its public versions, requires the front line of NATO conventional forces to stand firm and immobile under attack. It would not cede an inch. Nor would it move forward into enemy territory, lest NATO's defense posture be misread. Locked into position, NATO forces would absorb the entire impact of a Warsaw Pact attack. They would then have to bring it to a halt and defeat enemy units on the spot. Even with heavy numerical superiority, which NATO docs not have, success in carrying out such a strategy would be difficult. The doctrine therefore contributes to the general belief that conventional defense of Western Europe is implausible. The stage is thus set for possible early use of nuclear weapons by the West.

A forward defense posture is politically vital and remains valid. The Federal Republic could not support an alliance whose strategy called for surrender of large portions of West German territory even temporarily. But although longstanding political reservations have prevented active consideration of defense through mobile counterattack, these should be overruled to permit study of the advantages and disadvantages of such a doctrine. Under this concept, NATO would hold its armored forces in reserve behind a screen of defensive forces. The screen would include the additional West German reserve divisions, as well as British, Belgian, Dutch, and U.S. units already in forward position. They would absorb the impact of a first Warsaw Pact attack. The mobile armored forces would then counterattack, carrying the conflict into enemy territory. The counterattack would have the limited objective of encircling and cutting off the attacking force from its reinforcements in order to bring about a negotiated end to the conflict.

Following this concept in the event of conflict would not involve the massive destruction and civilian casualties in West Germany that many envisage. Much of the combat action would take place in the border area between West and East Germany and would involve direct confrontation of mobile military units. Indeed, aerial bombing of cities caused most World War II civilian casualties in West Germany. The Soviet Union has inadequate aircraft and would in any event be unlikely to make such bombing part of a conventional attack designed to capture West Germany's industrial plant intact.

It is not the object of this description of alternative possibilities for improvement of NATO conventional forces to argue for a particular version. Rather it is simply to make the point that with a Warsaw Pact challenge of finite dimensions and with the large standing forces and resources already at its disposal, NATO should be able to develop an effective and credible strategy for conventional defense that will—and this is the central issue—permit decreased reliance on extended nuclear deterrence.

In June 1982 President Reagan urged renewed attention to the Mutual and Balanced Force Reduction (MBFR) talks in Vienna. And in July 1982 Western participants in the negotiations presented new proposals.

There are strong reasons for seeking an Fast- West agreement in Vienna apart from the favorable impact it would have on the direction of public opinion in Western Europe toward NATO defense. The East–West military balance in central Europe is the largest and most expensive peacetime confrontation of manpower and firepower in human history. A military confrontation of these dimensions, with about one million men on each side heavily armed with nearly the whole range of conventional and nuclear weapons, entails a considerable risk that conflict will erupt through misperception or miscalculation of the military activities of the potential adversary. It is therefore imperative to press for an East–West arms control agreement aimed at stabilizing and defusing this confrontation. An agreement in Vienna could bring the West an increment of security at no cost. It could reduce and limit Soviet forces in central Europe. Additional measures could commit each side to notify the other of force concentrations or of movements of large numbers of troops. The two sides could also establish an annual quota of ground and air inspections to verify adherence to manpower ceilings and to increase advance warning of possible preparation for attack.

In the context of the other approaches discussed. Such an increment of security would legitimately make conventional defense appear more feasible to the West European public. In turn, public support for a continuing program to improve conventional forces in Western Europe might increase. An agreement along the lines foreseen by the NATO participants in the Vienna talks does not envisage reduction of armaments in an initial agreement. It would not prevent, therefore, continued improvement in NATO conventional forces. Nor is there anything inconsistent about simultaneously improving the equipment or logistical base of conventional forces while reducing the manpower of those forces through an arms control agreement with the Warsaw Pact states. The objective of both actions is to stabilize the East–West military confrontation in central Europe and to reduce the possibility that the confrontation could lead to war.

The two sides have already made considerable progress toward East–West agreement along these lines. The East earlier this year' presented a draft agreement incorporating many of the Western proposals. Some important aspects of a possible agreement remain in dispute: The two sides have not agreed on the actual number of Warsaw Pact ground and air force personnel now in the central European area. The East has not yet accepted the Western proposal for an annual quota of inspections of forces of the opposite alliance. Yet if President Reagan and the leaders of the other NATO states participating maintain the interest and involvement they have recently shown in the MBFR talks, they should be able, with time, to elicit similar involvement on the part of the Soviet leadership. It should then be possible to solve the data problem and to move forward toward a first agreement in Vienna. Such an agreement would not represent a zero-sum game benefiting only the West. Both East and West would benefit from a reduction in the risk of war in Europe and in the connected risk of nuclear escalation.

3.5 Defusing Nuclear Fears

As noted, the debate over the first use strategy is only one aspect of an underlying problem of NATO strategy—continued primary reliance on the American extended deterrent at a time when the United States has lost its nuclear superiority over the Soviet Union. In these circumstances, it is understandable that governments attempt to forestall change by trying to preserve appearances. But unless it is assumed that the United States can regain unmistakable and enduring strategic nuclear superiority over the Soviet Union, no amount of patching can restore the American nuclear deterrent to its former effectiveness or credibility. Attempts to do so primarily by improving NATO's nuclear arsenal in Western Europe do not have the right emphasis. They elicit controversy and compound the problem. The conclusion seems clear. The changed U.S.–Soviet nuclear relationship necessitates a change in NATO strategy, a shift away from primary reliance on nuclear deterrence to primary deterrence by conventional farces.

If NATO can implement such a revision of its defense strategy, then the issue of first use will become less significant. In the long run, depending on the evolution of West European views on the subject, NATO leaders could either drop the concept or allow it to atrophy, wrapped in increasingly restrictive provisos. In 2 or 3 years, West European opinion may realize more clearly the connection between something it dislikes—overemphasis on nuclear deterrence and nuclear armaments—and something many West Europeans still cherish as a symbol of American support for the defense of Western Europe—the first use strategy. In the long-run, West Europeans themselves may urge relinquishing that strategy.

Ultimately, when West European opinion is ready, a revised NATO strategy will assign nuclear weapons to the only role that over the longer run public opinion in Western Europe and the United States will support: deterring the use of nuclear weapons by the Soviet Union. To implement this more limited but still vital deterrent function, up-to-date U.S. nuclear weapons of various types will still have to be dedicated to the defense of Western Europe. Many such weapons will have to be deployed there, although many short-range tactical weapons could be withdrawn, preferably on a mutual basis through negotiation with the Warsaw Pact. Continuing deployment of these American nuclear weapons will itself represent a considerable deterrent to aggressive attack. But the revised strategy would place primary reliance for deterrence of conflict on conventional forces and supplementary, secondary, emphasis on nuclear deterrence.

A program that sought to decrease emphasis on nuclear deterrence by improving conventional forces and by moving to achieve a first MBFR agreement would greatly facilitate the creation of the new coalition of support for defense that circumstances in Western Europe call for. Such a program could help defuse the emotional debate in Western Europe over nuclear armaments and relieve political pressure during the long period that may be required to achieve agreement in the U.S.–Soviet negotiations on intermediate nuclear forces and those on strategic nuclear forces.

Ambassador Jonathan Dean (*center*) in a conversation with Prof. Dr. Karl Kaiser (to the *right*) and Dr. Dieter Dettke during a conference organized by the Social Democratic party in 1977. *Source* Personal photo collection of Ms. Theodora Schiro, daughter of Jonathan Dean. Reprinted with permission

Chapter 4
MBFR: From Apathy to Accord (1983)

Conventional arms control is the underprivileged stepchild of the arms control field, far overshadowed by the dramatic and controversial nuclear issue.[1] That governments and publics the world over focus their attention on efforts to regulate nuclear armaments is understandable given that nuclear war could obliterate entire countries and perhaps the human species. Nonetheless, this one-sided attention to nuclear questions overlooks the fact that the risk of conventional conflict between East and West somewhere in the world is greater than that of a conflict beginning with all-out strategic attack and that the most probable road to nuclear conflict between the United States and the Soviet Union lies through conventional conflict and its escalation to the use of nuclear armaments. In addition, conventional forces are far more costly than nuclear ones—on a ratio of at least five to one. In terms of saving money and decreasing the risk of war, including nuclear war, therefore, conventional arms control merits far more attention than it receives.

Nowhere is this more so than in the NATO-Warsaw Pact military confrontation in Europe. For a decade now, since the opening of the preparatory talks in January 1973, representatives of NATO and the Warsaw Pact have been negotiating in Vienna on reduction of their armed forces in Central Europe. Western negotiators call these talks the negotiations for *Mutual and Balanced Force Reductions* (MBFR); the official East–West title is "Negotiations on the Mutual Reduction of Forces and Armaments and Associated Measures in Central Europe." The MBFR negotiations cover active duty ground and air forces stationed in the territories of the Federal Republic of Germany, Belgium, the Netherlands, Luxembourg, the German Democratic Republic, Czechoslovakia, and Poland. Crammed into this

[1] This text was first published as: "MBFR: From Apathy to Accord", in: *International Security*, 7, 4 (Spring 1983): 116–139. The permission to republish this text was granted on behalf of MIT Press by Mr. Nick Yeaton, Permissions Assistant, The MIT Press, Cambridge, Mass. on 25 March 2014.

H. G. Brauch and T. Grimwood (eds.), *Jonathan Dean*,
SpringerBriefs on Pioneers in Science and Practice 19,
DOI: 10.1007/978-3-319-06662-2_4, © The Author(s) 2014

Central European area, which saw the beginning of the two most destructive wars in human history, is the largest and most potent peacetime military confrontation in the history of the world. On each side, about one million soldiers and airmen heavily equipped with modern conventional and nuclear armaments face one another. At this time, NATO ground and air manpower in the area including French forces—France does not participate in the MBFR negotiations—numbers about 990,000, and Warsaw Pact manpower about 1,180,000.

A confrontation of such enormous scope entails a built-in risk of human error, misperception, and miscalculation, which could lead to armed conflict. And East–West conflict in Central Europe could rapidly escalate to nuclear war, including all-out strategic strikes on the United States and the Soviet Union. Indeed, NATO strategy, heavily—indeed, overly—dependent on nuclear deterrence, explicitly foresees possible resort to nuclear weapons in the event of an overwhelming conventional attack by Warsaw Pact forces. And, although the Soviet Union has formally abjured first use of nuclear armaments, its forces in Central Europe and in adjoining areas of the Soviet Union are nevertheless equipped with a full spectrum of tactical and intermediate range nuclear weapons which they might use at any point in the course of a war in Europe.

The apparent stability of the huge Central European confrontation and its long continuation without outright conflict have tended to dull public attention to its risks. But all are living on an active volcano which could erupt at any moment. These risks and their potential consequences are so serious for the inhabitants of Europe, the Soviet Union, and North America that they justify the most energetic efforts at arms control.

Not only the risks, but the actual costs of the confrontation in Central Europe are immense. In the early 1980s, the United Nations estimated the world expenditure for armed forces at over $500 billion annually. About half of this sum is spent on NATO and Warsaw Pact forces in Central Europe. The United States spends roughly 50% and the Soviet Union about 60% of their total military budgets on these forces. The NATO countries with forces in the area—the Federal Republic of Germany, United Kingdom, Canada, and the Benelux countries—are also spending about 60% of their combined military expenditures on these forces. The Warsaw Pact countries located in the MBFR n-duction area—the German Democratic Republic, Czechoslovakia, and Poland—are spending about 95% of their total estimated annual outlay on armed forces on their ground and air forces in Central Europe, for a total, from both alliances, of nearly $250 billion annually.

Even in the fortunate event that it continues to be possible to avoid conflict in Central Europe, the military confrontation there will continue indefinitely. Unless there are really fundamental changes in the Soviet political system—and there is no evidence now that such changes are in the offing—the Soviet Union will keep large forces in a forward position in Central Europe. And NATO will need to maintain adequate forces to counterbalance them. It will remain the first priority of American foreign policy to deny Western Europe to the Soviet Union, to prevent dominant Soviet influence over the human and material resources of Western Europe, whether by military attack or by intimidation based on Soviet military strength. Such influence could tip the world balance of power in favor of the Soviet Union more decisively than increase of Soviet influence in any other area of the world. This consideration requires the continuing participation of American forces in the defense of Europe and justifies their cost to the United States.

Not only will the confrontation continue, but it is likely to become still more risky and more costly. Risks from the confrontation would become greater in a situation of increased East–West tension, of U.S.-Soviet confrontation in third areas, or of growing instability in Eastern Europe. Both sides continue to modernize and improve their forces in Central Europe, and risks of human error and misperception seem to multiply with increasing complexity of armaments. Furthermore, the confrontation will become even more expensive for both sides through the effects of inflation and the mounting costs of manpower and more sophisticated armaments in a period of low economic growth. In the West, the cost of modern fighter bombers is from $20–40 million each. The U.S. Army's Ml tank now costs up to $2 1/2 million. Although the Soviet Union has the advantage of a directed economy as regards production costs of its military equipment, like the U.S. it faces similar problems of increasing complexity and inflationary pressures in a period of declining economic growth.

Further problems for both sides make reaching an MBFR agreement even more desirable. Both NATO and the Pact will be encountering increasingly severe manpower shortages in coming years, especially in the Soviet Union and in Western Germany, where the possibility of lengthening the period of conscription, drafting foreigners residing there, and many other politically difficult expedients are being discussed. The United States, with a highly paid volunteer army, has increasing demands on its military manpower for units assigned to missions outside of Europe, like the Rapid Deployment Force. Under the impact of renewed U.S.-Soviet tension and widespread controversy over deployment of American Pershing II missiles and cruise missiles in Western Europe, there has been heavy erosion in the European defense consensus which has provided parliamentary support for NATO's effort. European anti-nuclear demonstrations, revived national feeling, and economic pressures in the United States have led to the reemergence of calls for the reduction of American forces in Europe, expressed for example in a September 1982 Senate vote. There is ample evidence of even more acute difficult)' and disaffection in Eastern Europe, where economic decline has undermined already weak public support for governments which are now being subsidized by the Soviet Union at a cost of over $20 billion a year.

The Vienna talks were designed to deal with problems of this kind. The immediate objective of the Nixon Administration and the allied European governments in entering the negotiations was twofold: to cope with pressures in the United States Senate for unilateral withdrawal of U.S. forces from Europe, and to control pressures in the European countries, especially the Benelux countries and the Federal Republic of Germany, to reduce the level of their own military personnel. The participating Western governments were successful in this immediate objective. But beyond it, especially in the governments of the Western European participating states, was the more far- reaching effort to see whether it would be possible to damp down the upward spiral of force increases on both sides with their associated increasing costs and risks, and to stabilize the Central European confrontation.

4.1 Initial Positions of the Two Sides

Since the underlying problems which motivated each side to begin the Vienna talks have continued, and will, in all likelihood, intensify still further, what has been done in the negotiations to deal with them?

The MBFR negotiations proper began in late October 1973. The NATO countries participating in the talks include seven Western "direct participants" with forces in Central Europe: Belgium, Canada, the Federal Republic, the Netherlands, Luxembourg, the United Kingdom, and the United States. Five NATO flank states, members of the alliance who do not have forces in Central Europe but whose territories abut on or are close to Warsaw Pact territory, are "special participants." These are Norway, Denmark, Italy, Greece, and Turkey. Of other members of NATO, Iceland has no armed forces, Portugal was in the throes of political change when the Vienna talks began, Spain not yet a member of the alliance, and France declined to participate, despite repeated efforts by both NATO and Warsaw Pact countries to persuade it to do so. All Pact members participate. There are four Pact direct participants—the Soviet Union, Czechoslovakia, the German Democratic Republic, and Poland—with forces in Central Europe. Hungary, Bulgaria, and Rumania are special participants.

From the outset of the main negotiations, the approach of the Western participants has had two main components: to reduce the active duty ground force manpower of both alliances in the Central European reduction area to a common ceiling of 700,000 men for each side, and to implement a package of "associated measures." NATO participants selected the objective of reducing the ground force manpower because, in their eyes, the most probable source of NATO-Warsaw Pact conflict would be attack by Pact conventional forces aimed at exploiting their ground superiority. (Both alliances agreed in the preparatory talks to exclude naval forces from the scope of the negotiations to avoid further complexity.) Given rough parity in air force manpower in Central Europe, it was considered that, if neither side had superiority in the level of its standing ground forces there, neither could count on an advantage in the crucial first days of a possible conflict. Consequently, the possibility that either side would decide to launch an attack would be reduced.

The associated measures which NATO has proposed are intended largely to verify compliance with residual ceilings after manpower reductions have taken place, and to reduce risks of misperception and miscalculation. This would be done mainly by providing prenotification of large force movements in and around the reduction area and the presence of observers to affirm the accuracy of these notifications, giving additional warning of force activities which depart from the normal pattern.

The initial Warsaw Pact approach to the negotiations, however, was quite different. Essentially, it was to freeze each national component of the forces of each alliance at its level at the time and to reduce the personnel and armaments of each of these national forces by the same across-the-board percentage—about 15%. NATO participants rejected this approach because the Pact forces are currently superior in numbers of military personnel and most major armaments. An across-the-board percentage cut in the various national components would therefore have involved their enduring acceptance of these numerical superiorities in treaty form.

The East's reduction approach also would have placed individual ceilings on each of the individual national forces making up the total forces of each alliance including, most prominently, West German forces. NATO participants rejected this approach also. They pointed out that national ceilings would mean that, if one NATO ally reduced its forces after an agreement was concluded, other allies could not make up the shortfall. Consequently, NATO participants insisted that manpower limitations be collective and be applied to the total forces of each *alliance* in order to permit internal adjustment by alliance decision, and that the only exception should be the forces in the area of the two world powers, the Soviet Union and the United States, which should be specifically limited.

4.2 The Changing Course of the Negotiations

The main thread of development in the MBFR negotiations has been slow but definite movement by the Soviet Union and other Warsaw Pact participants toward the Western concept of the common ceiling in ground force manpower and the associated measures. But, before describing the present status of the negotiations, I should note that the Western position, while retaining its central two elements, has changed in some respects during the course of the talks.

At the outset, Western, especially West European, apprehensions over the prospect of a force reduction agreement with the Soviet Union were considerable. In part owing to these feelings, and in part to give priority to U.S. reductions in the face of pressures for unilateral American withdrawals, Western representatives in Vienna proposed that reductions to the common ceiling be achieved through two separate, consecutively negotiated agreements. In the first of these projected agreements, the USSR and the U.S. would agree to reduce their forces. Soviet force reductions in this first phase would provide the basis of confidence for subsequent reductions by Western European members of NATO. The initial Warsaw Pact proposal, on the other hand, provided for a single agreement to be implemented in stages.

One drawback of the Western approach was that it provided no clear solution to the problem of the obligations to be undertaken in the first agreement by the

Western European direct participants and Canada with regard to limiting and reducing their forces. Soviet negotiators not unreasonably pointed out that the Soviet Union could scarcely be expected to reduce and limit its ground forces in Central Europe in return for U.S. reductions in a first phase agreement while leaving these other Western direct participants in a position to increase their forces at will and without any clear obligation to reach agreement to reduce their forces in the second phase of talks. Much of the time in the Vienna negotiations has been taken up with this issue With their latest proposals, made in July 1982, the Western participants have moved to accept the original Warsaw Pact concept of a single agreement on reductions with successive stages of implementation, in the first of which the U.S. would reduce 13,000 and the Soviets 30,000 men. This problem is therefore on the way to solution.

The West's most recent proposals also contain no provision for reduction of armaments. In 1973, East and West agreed to discuss reduction of armaments as well as military manpower. In 1975, the Western participants offered to withdraw a sizeable number of American Pershing I missiles, nuclear- capable aircraft, and nuclear warheads, and to limit their residual level in return for the East's agreement to withdraw a Soviet tank army and to make the further reductions needed to reach a ground force common ceiling of men on the basis of agreed-upon data. The offer was withdrawn 4 years later, in December 1979, when the East failed to agree to these conditions. The American warheads were withdrawn unilaterally as part of the internal agreement among NATO members to deploy modernized Pershings of increased range as well as cruise missiles. At the same time, NATO members decided to make the new' Pershing II missile the subject of negotiation on intermediate-range missiles in general. If accepted, the West's present MBFR proposals would mean that the issue of armament reductions would have to be handled in a successor negotiation.

The West made its July 1982 proposals in the form of a draft agreement, the major features of which have already been noted: to conflate the previously proposed two phases of negotiation into one phase and to include in the proposed single agreement assurances on manpower reductions by Western European direct participants and on a freeze of their forces between stages.

In 1979, the West had made an important expansion of earlier proposals for associated measures and these have been included without major change in its latest proposals. These measures included provisions for prenotification of entry of large numbers of troops into the reduction area and of major out-of-garrison activities; restriction of entry to and departure from the reduction area only through permanent exit-entry points manned by observers of the opposite alliance; periodic exchanges of force data; and, most important, an annual quota of ground and/or air inspections by each alliance of the other's forces.

Considerable progress has been made in the Vienna talks toward acceptance of the basic Western approach. In its draft agreement of February 1982, the East confirmed its earlier informal acceptance of the West's concept that the U.S. and Soviet Union should reduce first, that personnel reductions should focus on active duty ground force manpower, that there should be no obligatory reduction of air force personnel, that ground force reductions should be to an equal ceiling for both alliances, that this should be set at men for each alliance, and that, in principle, the ceiling should be collective for each alliance.

As regards associated measures, the East has agreed in principle, although with significant difference of detail, to temporary exit points with observers, prenotification of major movements into the area, prenotification of maneuvers inside the reduction area, and exchange of data after reductions. It has also agreed in principle to Western proposals for non-interference with national technical means and to establishment of a consultative mechanism to oversee implementation of an agreement. It has agreed to a Western proposal that neither the U.S. nor the Soviet Union would deploy forces withdrawn by agreement in a way which could increase the threat to the security of flank participants on each side, i.e., station the redeployed forces along the borders of flank states.

All in all, the East has tentatively agreed to about one-half of the individual items or propositions which would form the content of an accord along the lines of the Western draft agreement of July 1982.

4.3 Obstacles to Agreement

With strong reasons for moving to conclude an agreement and with considerable progress in the talks themselves, why have the MBFR talks hung fire for so long? Perhaps the underlying reason is that referred to at the outset—the absence of compelling drama in the subject matter of conventional arms control. Rational considerations—the cost and risk of the confrontation in Central Europe and the political costs of breaking off the talks—have kept the participating countries engaged in the Vienna talks. But the sustained high-level political interest necessary' to overcome the entrenched institutional interests, both political and military, of a two million-man military confrontation which has lasted for a generation has not been forth-coming on either side. It has been a vicious circle: lack of progress has diminished political interest and lack of interest has reduced the chances for progress. The MBFR talks have gradually moved down to the level of public and political attention given to such valuable but unsensational negotiating forums as the United Nations standing Committee on Disarmament in Geneva.

In practical terms, the MBFR talks have long been stalled on the data issue—the disagreement between East and West over the number of Warsaw Pact ground and air force personnel in the Central European reduction area, a difference of about 160,000 men for ground forces alone. The West's total figure for Warsaw Pact ground force manpower is about 960,000, while the total for the same forces provided by Warsaw Pact representatives in Vienna is about 800,000.

The need for agreement on data is intrinsic to reducing any element of armed forces by treaty. To be meaningful, a reduction must be accompanied by a limitation or ceiling on the remaining elements of the type being reduced, whether the element is artillery pieces or soldiers. Reductions without residual ceilings are illusory, and, to be effective, ceilings must be expressed in finite numbers. The only practical way to have assurance that the common ceilings on ground and air force manpower which are the core of the Western MBFR approach are being complied with is to check the size of the various national forces comprising these general totals. Since it is impractical to count all of the force personnel as of a given moment in time, the necessary tools for checking their size are agreement on the size of these forces pnor to reductions, backed by continuing data exchange and an adequate verification regime. For the forces of NATO participants, usable quantitative information on armed force strength is available through the parliamentary budgetary process. But this information is not available in the closed societies of the Warsaw Pact and has had to be assembled in bits and pieces by Western intelligence agencies. Over the years of the MBFR talks, these agencies have done a professional job of refining their estimates to respectable precision.

In the early period of the MBFR talks, prior to accepting the Western concept of reducing forces to a common ceiling, Eastern representatives in Vienna argued that for the East to accept this idea of reducing Warsaw Pact ground force manpower to a common ceiling of 700,000 men would require Eastern reductions significantly larger than Western reductions, and therefore would be disadvantageous to the East. At the same time, they refused to accept Western data as accurate, but also refused to submit data of their own. In late 1975-early 1976, Eastern representatives introduced a new line of argument: that Western armed forces in the MBFR reduction area had a large number of civilian employees whose functions were performed by active duty soldiers in Warsaw Pact forces in the area, and that the East should be consequently permitted to deduct these soldiers in counting its active duty military personnel. Then, in June of 1976, the East finally presented data on its ground and air personnel in the reduction area. Although the East claimed its figures were complete, and covered all personnel, including the service personnel Eastern representatives had earlier argued should be subtracted from their totals, their ground force total was at that time about 150,000 lower than Western figures for the same personnel, and the air force total about 20,000 short. Shortfalls of similar dimensions have been repeated in somewhat more detailed Eastern data subsequently presented.

Although the East claims that it applied the same criterion as the West of counting all full-time active duty ground and air force personnel in the reduction area, it would appear that some categories of personnel considered by the West to be active duty military personnel have been omitted from Eastern figures. Indeed, Soviet officials outside the talks have continued to make the argument, while discussing the data issue, that the Western armed forces have more civilian employees. In fact, East–West discussion in Vienna has identified some categories of Eastern European personnel which had been omitted. Western participants do not yet know what specific groups may have been omitted from the figures on Soviet forces in the reduction area. The straightforward way to resolve these differences would be to compare Western and Eastern figures for the same components of Soviet forces, in order to identify the force components where the difference is greatest and to discuss these in more detail. But Eastern representatives have been unwilling to provide the more detailed figures on Soviet forces needed to engage in this comparison, claiming that it would reveal sensitive intelligence information. Instead, they have argued with increasing assertiveness that agreement on force data is not a prerequisite for a force reduction agreement, claiming that if ground forces in the area were reduced to an agreed-upon common ceiling of 700,000 men for ground forces and the ceiling were verified, there would be no need either to agree on force levels prior to reductions or to specify the exact size of reductions.

The difficult of the data dispute is compounded by the fact that, since the dispute first emerged, the Eastern participants have accepted the Western concept of reducing ground force personnel in the area from their level at the time of concluding an agreement to a common ceiling of 700,000 men. According to Western data, this would require a reduction of some 260,000 Warsaw Pact ground force personnel in return for the West's reduction of approximately 90,000. Since the Soviet Union makes up one-half of Pact forces in the reduction area, their share of Eastern reductions would be about men instead of the roughly 50,000 men which the Soviets would be expected to reduce according to Eastern data. (The Soviets also argue that their unilateral reduction from East Germany of 20,000 men in 1979–1980 should be subtracted from their total reduction.) It is possible that reductions of the size shown in Western data are not acceptable to the Soviet Union, particularly in a time of heightened concern about political stability in Eastern Europe focusing on Poland. That is, the issue of the size of Soviet and other Pact reductions they would have to agree to if they agreed that Western figures on Pact forces were correct may underlie Soviet unwillingness to move to resolve the data disagreement.

The pattern of Soviet behavior in the Vienna talks is one of cumulative concessions to the Western position in other respects while determinedly withholding practical cooperation (i.e., providing detailed figures) on the data issue. This behavior indicates that very important interests underlie the Soviet position on data. It is not completely clear what those interests are. The possible motives include: loss of prestige if figures different from those presented by the East were to be agreed upon; deliberate use of the data issue by some group or groups in the Soviet Union to block an agreement; and, as noted above, concern that reductions called for under Western data are too great to absorb.

The Soviets have shown strong interest in an initial Soviet-American force cut, giving convincing indications of willingness to make force reductions of limited size. In fact, in the last months of 1979 and the first half of 1980, they unilaterally withdrew some 20,000 men from East Germany, using withdrawals previously planned for MBFR in an unsuccessful effort to persuade NATO countries not to decide to deploy new American intermediate-range missiles. Soviet representatives have repeatedly argued in Vienna that the size of total Eastern cuts shown in Western data was disproportionately large when compared to Western cuts. But Soviet representatives have not argued explicitly that total reductions of the size needed to reach the common ceiling as shown by Western data were too large for them to absorb, nor have they directly addressed the issue of the size of Pact reductions as such, or, since 1976, attempted to argue that they should be permitted more forces in the area than the West. Instead, they have apparently preferred to carry out the discussion about the size of reductions from behind the bulwark of an unconvincing and uncooperative position on the data issue, an approach which has done considerable harm to the international reputation of the Soviet Union as well as effectively blocking progress in the MBFR talks.

It is no secret in Vienna that the Soviet Defense Ministry has the final say on release of data on Soviet forces and, in practice, on data for the other Warsaw Pact military forces, and that such data are considered security information. The Ministry prepared data on Pact forces presented in Vienna both originally and subsequently, and seems to have rejected several suggestions that the Soviet Union provide more data. This points to one of the possible explanations for the data problem: namely that the Soviet military' establishment would prefer not to see a contractual East–West force reduction agreement with the West—whether because of the size of the projected reductions, or because of concerns about long-term damage from such an agreement to the rationale for the ground force claim on allocation of Soviet resources (a reduced Western threat means a reduced need for this expensive component of Soviet forces), or because of concerns about the effect of Soviet troop withdrawals on political stability in Eastern Europe If this military resistance is the source of the problem, it would also appear that the Soviet political leadership has thus far not been willing to assume the political costs of the friction which would come from overruling the military and breaking its hold on Soviet military data and the Soviet position on the Vienna talks.

A second area of practical difficulty in the Vienna talks—and it is probably closely related to Pact views on the advantages of military secrecy—is Warsaw Pact reluctance thus far to agree to a continuing role for observers or inspection of personnel in the context of Western proposals for associated measures. Although Pact representatives have agreed to check points with observers to check the departure of military personnel withdrawn from the reduction area by agreement, they have not thus far agreed to the permanent establishment of exit-entry points with observers of the other side through which military personnel of outside countries with forces in Central Europe would be required to enter and leave the area. Nor have they agreed to the obligatory presence of observers at out-of-garrison movements of forces over a specified size who could assure that such movements did not form part of a preparation for attack. And they have not yet agreed to an annual quota of ground and air inspection of the forces of the opposing alliance which could provide valuable help in verifying manpower units and in gaining early warning of unannounced force concentrations. Here, too, the main area of resistance thus far has been the Pact military establishment.

Western security interests have led Western participants to place greatest emphasis on reduction and limitations of Soviet forces as compared to other Pact forces. They have argued that the only legitimate restriction on the common ceiling is a sublimit on the forces in Central Europe of the two world powers. Given the fact that the West German Bundeswehr is today possibly the best single force of those in Central Europe, it is not surprising that Warsaw Pact participants in turn have placed greatest stress on reducing and limiting these forces. Although accepting the Western approach that manpower limitations be collective for each alliance. Pact negotiators have continued to seek agreement to some restriction on the size of West German forces, arguing that the Western concept of collective ceilings would permit the Bundeswehr to increase in size if other Western allies decreased the number of their forces in Central Europe. But sizeable German increases are highly improbable, not least because of manpower shortages, and the NATO allies must in any event be in a position under an agreement to maintain their forces at the permitted limit. Warsaw Pact concerns on this subject, whatever their legitimacy, are outweighed by the fact that Soviet forces in the nearby USSR would not be under any limit.

Although Soviet behavior on the data problem is the greatest practical difficulty of the Vienna talks, there have also been difficulties on the Western side. High level political follow-through on the Western side has been sporadic. Successive United States administrations have given the MBFR talks lower priority than nuclear arms negotiations with the Soviet Union, considering, inaccurately, that the latter negotiations had a more direct bearing on national security. Once domestic pressures for American troop withdrawals from Europe subsided, American political leaders have seemed to consider the MBFR talks as primarily a West European interest and consequently not to devote consistent attention to them.

Senior American military officers have shown a fairly positive attitude toward the talks and an awareness of their potential for stabilizing the confrontation in Central Europe, and military officers assigned to work on the negotiations directly have given valuable help. But the absence of consistent political level interest in moving to an outcome has often left the field to institutional interests of one military service or another, with resultant damage to the realism of the Western negotiating position. For example, during one period of several months when Soviet interest in reaching an agreement appeared especially strong, it was impossible to obtain a decision on the specific number of American troops which would be reduced in the first phase of a simplified reduction approach, even though presenting a specific amount for American reductions was an elementary requirement for reaching an agreement and for showing interest in doing so. And, though East and West have long agreed that first-stage Soviet and American force reductions should be preceded by a list of the units to be reduced by each of these two armies, it has not been possible for a decade for the U.S. to present even an illustrative list of the types of units which it would be withdrawing from Europe in a first stage if an agreement were concluded.

In turn, some of the problems which have arisen during the talks have involved the West European participants. West Germany was originally the moving force behind the MBFR concept and logically should have been the locomotive pulling the enterprise forward. But, although at the outset it was so enthusiastic about the talks as to insist for a time that its forces be among those reduced in the first stage, it became less active as divisions of opinion emerged within the German government on the subject matter of the talks. This development has had consequences for the Western position on practical issues as well as for the degree of political impetus given the talks. On the phasing issue, it took a long time for the NATO allies to work out ways of meeting the East's reasonable insistence on having real assurance that West Germany and other Western participants would actually make a sizeable reduction of their forces if the Soviet Union reduced first. At times, some German leaders appeared more impressed by the possibility that West Germany, because it was located in the reduction area, might be undertaking obligations, such as acceptance of inspection of its forces, which its major European community partners—France, the United Kingdom, or Italy—would not be undertaking at home. Borrowing a French argument against the MBFR talks, they considered that such obligations could establish a politically invidious "special zone" in Central Europe. Such concerns are understandable, but they ignore the fact that the Benelux countries agreed to become part of the reduction area specifically to avoid a politically unpalatable emphasis on Germany, and they ignore the potential gains from such measures for the security of a Federal Republic which history and geography have placed in the most exposed position in the East–West confrontation. It is unclear what attitude a new West German government will have on these issues after elections in the spring of 1983, although it is clear that it will wish to continue its participation in the MBFR negotiations.

4.4 Advantages of an MBFR Agreement

An MBFR agreement along the lines under consideration in Vienna could make conventional war in Europe less likely and give both sides an increase in security at no additional cost. Staged reduction of the ground force manpower of both sides to an equal level of 700,000 men would mean that, as long as the agreement is being respected, neither side could increase its ground or air force manpower in Central Europe, either through reinforcements from outside or through mobilization of reserves, to the level needed to assure the *success* of an attack. Local superiority would be possible through using forces already in the area, but neither side would have the manpower superiority in the crucial initial stage of conflict which could give confidence of a successful attack. In practical terms, these ceilings would mean for the West that the Soviet Union could not augment its forces in Central Europe prior to an attack and that, if it did attack with minimum preparation and without reinforcement, it would be obliged to use Pact standing forces in Central Europe, for which standing NATO forces in the area would be a match The Warsaw Pact would have the same assurance as regards possible attack by NATO forces.

The associated measures proposed by the West would help in verifying that manpower ceilings were being respected and provide early warning of troop movements or concentrations departing from the normal pattern. Movement of large numbers of personnel into Central Europe would require prenotification and personnel entering or leaving the area would have to pass through permanent observer posts. Larger movements of forces within the area would have to be reported in advance, and obligatory observers would assure themselves of their character. Periodic exchanges of data would provide both sides with information on the strength of specific force components. This information could be checked by inspection teams on a sampling basis to assure compliance with manpower ceilings. Ground and low-flying air inspections could give all-weather capability to identify and check for unannounced force movements and concentrations. This capability would be an important addition to satellite photography of the type in service up to now, which is partially blinded by the clouds that cover Central Europe for 60 % of the year.

Taken together with information provided by national means, implementation of these measures would provide dependable assurance that neither side was preparing for conflict and would insulate the military confrontation in Central Europe against the destabilizing effects of accidental, unauthorized, or mistaken actions by individual units or personnel which might otherwise be misinterpreted and set in train a sequence of actions leading to conflict. The West has also proposed an East–West standing consultative committee which could exchange information, discuss ambiguous activities, and perform crisis management functions of the kind recently proposed by Senators Nunn and Jackson for the Soviet Union and the United States.

There is no guarantee that the provisions of an MBFR agreement would be respected by both sides for all time. But significant violations of contractual obligations, such as entry into the area of units of personnel without preannouncement, would be detected and would provide political authorities of the other side with unambiguous evidence of hostile intention. A number of observers have seen as a major weakness of Western defense the possible reluctance and slowness of NATO political authorities acting on the basis of ambiguous intelligence indications to make timely decisions to move NATO forces into forward defensive positions under circumstances of a possible pending Pact attack. Violation by Pact forces of clear treaty commitments would help NATO political authorities to reach decisions rapidly. The measures would have similar benefits for Warsaw Pact leaders, who might be confused by conflicting information and signals from the NATO countries, and thus help to avoid Pact decisions based on misperceptions.

4.5 Criticism of the Negotiations

Like every sizeable enterprise of considerable duration, the MBFR negotiations have attracted some recurrent criticisms. The Western negotiating concept has been criticized by some as irrelevant and not focused on the right subject matter. Given the dimensions of the military confrontation in Central Europe, it is not surprising that security and arms control experts have made many, often divergent, suggestions on how best to deal with it.

One recurrent criticism of the design of the Western position is its focus on reduction of the number of individual servicemen and its failure to provide for reduction concepts considered to have greater stabilizing effects, such as reduction of combat units or reduction of armaments, including tanks and nuclear armaments. To criticisms of this kind, it can be said that there are many versions of arms control, but no ideal arms control. The important security benefits for both sides from implementation of the present Western approach have been described. The focus on military manpower is justified by the fact that soldiers and airmen are the final agents of military conflict—any other force element can be eliminated, and conflict could still take place. Soldiers are different in quality, morale, and training, but as individuals they are more comparable to one another in military terms than military units of different sizes, organization, and equipment. Manpower limits would also cap the most expensive component of armed forces. It is logical and necessary to seek to reduce the huge accumulation of weapons in Central Europe at some point. But the Pact's numerical superiority in most major armaments and the drawbacks of accepting in an agreement an Eastern numerical superiority make it difficult to design a concept for reducing armaments in Central Europe which is both effective in the arms control sense and relatively negotiable. For example, reduction of tanks on both sides to parity at a lower level as with military manpower, a possible approach, would require much larger Pact than Western reductions. However, tanks and nuclear armaments in Central Europe have been part of the Western negotiating program in the past, and this subject should be addressed in future MBFR negotiations, after the reduction process has been validated by the implementation of a first MBFR agreement.

Other experts argue that it is wrong to concentrate on force reductions and that there should instead be agreement on a larger number of associated measures or confidence-building measures limiting the activities or location of existing forces. But there would be something disproportionate about limiting the activities of existing forces in Central Europe while making no effort to reduce their present enormous scope. Force reductions, even if not comprehensive, are evidence of seriousness of purpose. The interaction of residual ceilings and associated measures has a much greater stabilizing effect than associated measures alone. Furthermore, associated measures are no easier than force reductions to agree on within the Western coalition or between East and West, and they have no savings or cost-avoidance effect. In an agreement which is at least half completed, the major effort should be to bring it to completion. Armament reductions and further associated measures can be dealt with in a successor negotiation.

Another criticism of the Western MBFR concept is that the Central European reduction area is too circumscribed, in that it does not include at least the western portion of the USSR which abuts on the agreed reduction area. These critics point out that, whereas withdrawn American forces would have to go back across 3,000 miles of ocean, Soviet forces withdrawn under an MBFR agreement would only retreat a short distance to the USSR and probably the western part, from which they could rapidly return to Central Europe in the event of conflict, when any MBFR agreement would of course be nullified. However, these troops could not return more rapidly than the Federal Republic could mobilize its reserves of 750,000 men; and, moreover, Soviet units withdrawn pursuant to agreement would probably have been placed in a state of lower readiness. The more basic reply is that an MBFR agreement acceptable to both sides cannot be expected to provide either side absolute protection against any form of attack. In the ways described, it merely provides an additional increment of security to both sides.

Critics also point out that Soviet forces in the western Soviet Union would not be subject to any of the manpower limitations, reporting requirements, or inspections which would form part of an MBFR agreement and could increase. However, the NATO participants considered, when they agreed to the MBFR reduction area, that it was not feasible at that stage to seek Soviet agreement to limit forces on its own territory, and that the security gains from limiting Soviet forces in the forward position in Central Europe still outweighed these drawbacks. This remains the case.

The Western MBFR participants have also tried to deal with concerns like these by offering to include the territory of NATO flank participants in the area in which out-of-garrison activities would be notified in return for inclusion of a portion of the western USSR, but the Eastern response thus far has been negative. It should be kept in mind in any event that the number of active duty Soviet military personnel in units in the western USSR which would be used to reinforce those in Central Europe in the event of conflict are counterbalanced by roughly equal numbers of French forces. Moreover, with a few exceptions, these Soviet units are in a low state of readiness. This suggests that what might be desirable in a future MBFR negotiation is to explore whether some form of assurance might be provided on the readiness state of these forces in the western USSR, like prenotification of call-up or mobilization of reserves. But this could raise issues concerning notification of similar mobilization in Western Europe or the United States, which would be the main sources of reinforcements for NATO forces in Central Europe, the whole subject of which is too complicated to deal with in the present negotiations.

The refusal up to this time of France to join the MBFR negotiations is a defect of considerable dimensions. The exclusion of the nearly 50,000 men composing French ground forces in West Germany also creates practical negotiating problems, since they have to be taken account of in some way. The French motivation in refusing to join the MBFR talks is not entirely clear. Ostensibly, it is to prevent the creation of a special zone in Central Europe leading to the possible neutralization of West Germany. But this outcome, unlikely as it is, is more likely with France outside the negotiation than in it. Strong Gaullist skepticism about the desirability of arms control negotiations, resolved only late in the Giscard government, continued French rejection of participation in the NATO integrated military command, and consequent reluctance to meet the requirement for the NATO participants to have a unified position in Vienna seem more plausible explanations for French abstention.[2]

Another line of criticism of a passible MBFR agreement—one whose counterpart criticism is raised just as frequently among Warsaw Pact governments as in the West—is that conclusion of an MBFR agreement could lead to a state of unjustified euphoria in the West which could mislead public opinion as to the true nature of the threat to the Western coalition by the Soviet Union at this juncture, and undermine willingness in the West to support the large defense budgets which will continue to be necessary.

But, as has been true from the outset of the MBFR talks, when similar concerns were raised, the extent and direction of the public impact of an MBFR agreement would depend largely on how Western participating governments present the agreement to their publics. At best, an agreement would provide only for a limited reduction of Warsaw Pact military manpower. A large, well-equipped force, backed by the force in the nearby Soviet Union, would remain in Central Europe, and Western governments can make this clear to their citizens. Important as a first agreement along the lines sought by the West would be, it would not have reduced the large stocks of Warsaw Pact tanks and nuclear armaments in the reduction area. It would be a limited experimental beginning. Indeed, the final decisions by participating Western governments to proceed to an agreement could be combined with specific decisions by the same governments on force improvement.

4.6 Benefits Associated with an Agreement

The still broader political concerns about the general desirability of entering into any new arms control agreements with the Soviet Union which are raised from time to time in the United States and other Western countries have to be balanced against the costs of continuing the confrontation in Central Europe without an agreement as well as against the gains from achieving an agreement. The specific

[2] NATO countries participating in the MBFR talks also support the French proposal for a Conference on Disarmament in Europe, which has been under discussion in the Madrid CSCE review conference, together with the parallel Soviet proposal for an all-European Conference on Military Detente and Disarmament in Europe Tins possible conference would focus on confidence-building measures in a wider area than Central Europe, including the territory of France and the western USSR as well as of the European neutrals. There has been little effort to relate it to the MBFR negotiations 1/some version of a CDE conference is finally agreed on, the most logical relationship between the two negotiations is that of two concentric circles, with the MBFR talks reducing and limiting forces by treaty in the area of greatest concentration and the CDE covering a wider area with a looser network of obligations. If both negotiations conclude successfully, some way should be found of amalgamating them, rationalizing the process of conventional arms control in Europe.

gams for the security of both sides from a first agreement have already been outlined here. Conclusion of an agreement could lead to other collateral gams for the Western coalition. It was pointed out at the outset of this article that a main purpose of NATO's defensive forces is to prevent use of Soviet military power to intimidate Western European political leaders, to insulate the decisions of these leaders from concerns about the possible negative reactions of a militarily powerful neighboring Soviet Union. The reverse side of the concern about euphoria is that an operating MBFR agreement—withdrawal of a considerable number of Soviet forces, respect for a residual ceiling on remaining forces, and successful operation of a system of associated measures—would provide important reassurance for Western European political leaders against possible Soviet pressures based on military power. In its political effects, this reassurance would be the equivalent of a considerable Western force increase, but at no additional cost.

Conclusion of an MBFR agreement could also have a positive effect on defense morale in Western Europe. The support of Western European public opinion for the NATO defense posture has been severely undermined by controversy over the deployment of new American nuclear missiles in Europe. Peace demonstrations of surprising size and repeated public opinion polls evidence widespread public concern, especially in West Germany and the Benelux countries, that NATO's present defense posture relies too much on nuclear armaments. The widespread and correct conclusion is that NATO's conventional force posture should be improved to decrease reliance on nuclear armaments and on their possible first use at an early point in meeting a Pact attack. An approach combining conventional force improvement with achievement of an arms control agreement on conventional forces could make a program of improvement appear more feasible and move toward establishing in Western Europe the new consensus on defense which the present difficult situation calls for. It is not inconsistent to take simultaneous steps to improve the equipment or logistical base of NATO forces in Central Europe and to reduce the manpower of these forces through a reciprocal agreement with the Warsaw Pact states. The objective in each case is the same: to stabilize the East–West military confrontation in the center of Europe and to reduce the possibility that its existence could give rise to war.

For NATO participants, the savings from an MBFR agreement calling for reduction of about 90,000 soldiers would be distributed among the seven countries with forces in Central Europe. The reductions in manpower would make possible some savings and help to focus resources on improved armaments and infrastructure. But these actual savings will be of limited dimensions and are likely to be absorbed in the rising cost of military equipment. Nonetheless, an MBFR agreement could have an important long-term impact on defense costs. If Soviet leaders consider that the operation of an agreement has in fact stabilized the confrontation in Central Europe, they may feel in a position to cut back on the present rate of modernizing their forces. This would permit Western force planners to follow suit.

An MBFR agreement would have the additional long-term advantages of providing a framework for evolutionary change in Eastern Europe. Essentially an alliance-to-alliance agreement, it would provide assurance that Western countries accepted the continuing participation of the Eastern European countries in the Warsaw Pact and would not seek to take advantage of social and economic change in Eastern Europe by building up their forces in contiguous areas. It might thus make Soviet leaders less reluctant to permit such change, however grudgingly.

4.7 Will There be an Agreement?

Against the background of the development and problems of the MBFR negotiations over the past decade, what are the prospects that an agreement will actually be concluded?

Both sides in the MBFR talks have put forward draft agreements on generally parallel lines. Parts of their content are identical; parts are closely similar. Most remaining open issues have been discussed for years and the range of possible solutions has been thoroughly canvassed. Two main areas of difference remain—the data dispute and the reluctance of Pact participants to agree to reciprocal observation and inspection. There are indications that the latter view may be changing. If so, this will leave the data issue as the main open issue in the talks.

To resolve it and to settle the practical details of an agreement like troop lists, a sustained high-level political push from the Western participants will be necessary. Only this level of political engagement will elicit its counterpart on the side of the Soviet leadership, the engagement needed to override the resistance of the Soviet military establishment to decisive progress in the Vienna talks. If high-level political engagement can be achieved on both sides, completion of an agreement could follow within a year. Soviet concerns about the future of Poland are clearly a hindrance to moving ahead. But it is possible that, in circumstances where the West is showing high-level interest in concluding an agreement, the new Soviet political leadership may consider that gains from stabilizing the situation in Central Europe and establishing a better relationship with the United States and other countries of the Western coalition outweigh these concerns. Realization that an MBFR agreement is possibly the only East–West arms control agreement which can be completed within the next 2 years of Mr. Reagan's presidency may give both sides some urgency.

Technically, the East now owes the West a substantive response to the Western proposals of July 1982—at the least, if the East does not find these proposals acceptable in their entirety, in the form of a counterproposal. To be taken seriously in the West, such suggestions must tackle the data issue directly. The necessary first step toward solution of this issue is for the Soviet Union to clarify in direct terms whether the primary problem underlying this dispute is, first, genuine disagreement over the current number of Pact soldiers in the reduction area, based on different views of the actual count, or, second, the size of Warsaw Pact reductions, particularly Soviet reductions, which would be needed to come to an equal 700,000-man ceiling with the Western participants. If it is the first problem, then resolving it objectively should be possible. For example, the difference between Western figures for Soviet forces and the figures provided by the Soviets is only about 10 % of the Western total, and it should be possible to find the reason for this limited divergence. The straightforward method of comparison of the data of both sides for the same type of Soviet units could do the job. There are a number of other possible solutions. One of these would be short-notice inspection on a sampling basis of a number of Pact units selected by the West. This approach need not be unilateral or involve prestige considerations. The Warsaw Pact participants could also use inspections to clarify doubts they may have about details of data provided by Western direct participants on their own forces.

It is important to know the real nature of Soviet motives in the data dispute, for if the underlying problem is really the size of Eastern reductions, continued efforts to resolve the data issue as such will not deal with the real issue, and will not succeed. If the real issue underlying the data dispute is reductions larger than the Soviet Union considers possible to implement at this time, this issue has to be addressed directly. Several solutions are theoretically possible, subject of course to their ultimate acceptability to ail participating Western governments. One approach would be to redefine the group of personnel to be reduced to make total reductions on both sides smaller, such as a common ceiling on certain types of combat forces. Other theoretical possibilities would be to return to the West's 1979 concept of a limited first phase, or to protract and draw out the process of reductions to the common ceiling over a longer period than is now foreseen. A further, less desirable possibility would be to postpone reductions and precede them by a new first phase consisting of a freeze and associated measures which would be implemented while discussion of reductions and data continues. Because it has never been clear thus far that the main difficult underlying the data issue, if indeed it is, is the question of the size of Eastern reductions, such concepts have not been seriously studied in the West. Consequently, the reaction of Western governments to such approaches is quite uncertain.

The important thing now is to bring the Vienna negotiations to a first outcome on the basis of the existing structure already agreed upon by East and West. Future negotiations should deal with reduction of armaments in Central Europe, especially with shorter range nuclear weapons not dealt with in the Geneva INF talks. To be most effective, such reductions should in some way take account of armaments stationed in France and the western USSR. Future negotiations should also consider a follow-on group of associated measures, of which a major one should be agreement to move back military bridging equipment on both sides. But that is for later

Because sustained high-level political attention is so important to successful conclusion of a first MBFR agreement, it would be desirable to follow the recommendation of former British Foreign Secretary Owen, recently reiterated by the Palme Commission, to hold a meeting of the foreign ministers of participating countries in Vienna within a year to discuss a written report from the negotiators of progress in the talks summarizing points of agreement and disagreement. This step could give the negotiations the sustained impetus they need. Quite aside from all specific suggestions, proposals, and timetables, no result will be achieved in Vienna unless the heavy inertia of the institutional military interests which the Central European confrontation has created on both sides can be counterbalanced by top-level political interest in an outcome.

Whether or not the participating governments provide the impetus needed to conclude a first MBFR agreement within a few years, for the West, continuing efforts to reach such an agreement are an essential part of the necessary long-term effort to improve NATO conventional forces and to reduce the present, increasingly unsatisfactory reliance on nuclear deterrence. Serious efforts to reach an agreement will reassure Western publics that there can be some limit to the sacrifices they are being called on to make to improve conventional forces. It will reassure the Warsaw Pact countries that the improvement of Western conventional forces is not in the service of an aggressive long-term objective. For both East and West, the stakes in the future development of the vast military confrontation in Central Europe are immense. They include, on the one hand, the important security gains from an MBFR agreement. They include, on the other, the possible losses from failure to deal with the Central European confrontation through an arms control agreement. These losses range from the increased cost and risk to both sides, to the third of the world wars which have started in Central Europe. Realization that this war would be the last of the series will sooner or later generate the political push needed to bring the Vienna talks to success.

Chapter 5
Alternative Defense: Answer to NATO's Central Front Problems? (1987/1988)

In the early 1980s, intense controversy over the deployment of new American intermediate-range nuclear missiles brought insistent questioning in Western Europe of the existing NATO defense posture and its dependence on nuclear weapons, along with a steady stream of proposals calling for its radical reorganization.[1] These ideas have been most numerous in the Federal Republic of Germany (FRG), where the bulk of NATO ground forces face the main Warsaw Pact standing forces, and where opposition to missile deployment has been most intense. The resultant debate between the proponents of new approaches and the supporters of the current NATO force posture has had two main themes: the possibility of a defense based on a drastic cutback of NATO nuclear weapons, and discussion of the merits of *static defense* (based on high-technology microelectronics—sensors, and rocket-launchers using precision-guided munitions) versus *mobile defense* (based on tanks and other armored platforms).

In the context of a possible agreement on intermediate-range nuclear forces (INF) between the United States and the Soviet Union, Western military experts and diplomats are anxiously looking for new ways to improve NATO conventional defenses—and also for a productive negotiating position for the new NATO—Warsaw Pact 'Atlantic to the Urals' negotiations scheduled to open in the spring of 1988. Proponents of these new approaches claim to have answers to both needs. This article examines the contentions of those supporting the new proposals that their adoption would sharply reduce the dangers of the formidable force concentration in central Europe and, even if implemented unilaterally by NATO, would contribute to building down that confrontation.

[1] This text was first published as: "Alternative Defense: Answer to NATO's Central Front Problems?", in: *International Affairs*, 64,1 (Winter 1987/1988): 61–82. The permission to republish this text was granted on behalf of Blackwell Journals by Ms. Butler on 11 March 2014.

H. G. Brauch and T. Grimwood (eds.), *Jonathan Dean*,
SpringerBriefs on Pioneers in Science and Practice 19,
DOI: 10.1007/978-3-319-06662-2_5, © The Author(s) 2014

There is no widely accepted standard designation for these suggestions for reorienting NATO's defense posture. In the analysis that follows, they will be called proposals for *alternative defense.* But before turning to a detailed analysis of alternative defense proposals, it may be useful to place them in context against the overall setting of developments in NATO thought and criticisms of the NATO posture since the early 1980s.

5.1 Orthodox Criticisms of NATO's Force Posture

The controversy over deployment of the new American missiles in the early 1980s brought with it active public questioning in Europe both of NATO's existing defense posture and of the underlying reasons for the NATO-Warsaw Pact military confrontation itself—controversy heightened by the deadlock that had been reached in the arms control process, both in the Mutual and Balanced Force Reduction (MBFR) talks and the Strategic Arms Reduction (START) and INF negotiations which collapsed in 1983. These trends crystallized into fundamental criticism of NATO's defense posture, and led to the development of many proposals to unilaterally structure Western defense in new ways and on a non-nuclear basis. Because the NATO-Pact confrontation is at its peak in central Europe, most of these proposals were aimed at changing NATO posture in the FRG, and especially at changing the organization of the German armed forces, the Bundeswehr.

The extra-parliamentary protest movement in Western Europe declined markedly after deployment of the new missiles began in 1983–1984 without the much predicted culmination in East–West conflict. Yet the end of the dispute over INF deployment did not mark the end of public concern over NATO nuclear strategy, especially over its heavy dependence on possible early first use of nuclear weapons. As a result, the aims, objectives and proposals of the anti-nuclear movement became largely incorporated into the programs of the political parties of the left in the northern tier countries—in the United Kingdom, the FRG, the Netherlands, Denmark, and to a lesser extent Norway and Belgium. In the later 1980s the anti-nuclear argument has thus become a party political one.

Since the designation of Mikhail Gorbachev as General Secretary of the Soviet Communist Party in March 1985 simultaneously with the resumption of the US-Soviet arms control talks, there has followed a series of unprecedented Soviet concessions to the American arms control position, leading in December 1987 to the conclusion of an agreement to eliminate all Soviet and American INF missiles. But the prospect of an INF agreement has once again focused European attention on the conventional NATO-Pact balance—some argue that elimination of the American INF missiles would strip away a valuable counterweight to Pact numerical superiority in conventional forces and make this superiority more important—and on the prospects for building down the conventional confrontation in the new Atlantic-to- the Urals negotiations which begin in 1988.

Besides 'flexible response', NATO strategy in central Europe has for several decades been based on a second main concept—'forward defense' along the German border, with the aim of holding a Pact attack as far forward as possible. Forward defense too has come under criticism, but mainly from professional military officers who criticize its lack of strategic mobility—the result of forward defense's prohibition on surrendering ground to attack and its inhibition on deep counterattack into enemy territory, which it considers excessively 'offensive'. Yet given Germany's military and political importance in NATO, the concentration of German population and industrial capacity near the inner-German border and the deterrent effects of forward deployment of foreign allied forces, forward defense appears logical in the circumstances.

NATO ground forces called on to carry out forward defense are organized mainly in the form of mobile, heavily armored formations—main battle tanks, armored personnel carriers and armored self-propelled artillery and anti-aircraft weapons. Prevalent opinion in NATO circles considers this to be the optimum method of organizing an all-purpose force with maximum flexibility of use. The mobility and armor flanks of an armored force protect it against artillery or air attack at the outset of invasion and permit it to defend at points of actual enemy attack, to move to deal with possible enemy breakthroughs, and to retaliate against invasion through at least a limited degree of counterattack on enemy territory, engaging enemy columns on their more vulnerable flanks. An armored force also has the potential power to push an invading force back out of German territory.

Many Western military experts have faulted NATO forces organized around armored formations as insufficiently strong to carry out their task in relation to Pact ground forces, with too few active-duty ground force divisions, inadequate operational reserves to cope with breakthroughs, and insufficient staying power from low ammunition and fuel stockpiles. Their worries are intensified by the increasing costs of major equipment—over $2 million for a modern tank, over $40 million for a modern fighter-bomber. In the case of the FRG, the costs of social services are expected in the 1990s to cut increasingly into revenues available for defense. And as a result of the widespread use of the birth control pill in the 1970s, by 1994 a shortfall of some men per year is expected in Germany in each age group liable to military conscription. (By the end of the 1990s this shortfall could result in a decrease of about 200,000 from the present Bundeswehr strength of about 500,000 men. This gap could be made good only by expensive additional payments to volunteer military personnel, and then only in part.)

5.2 The Development of Alternative Defense

Proponents of alternative defense accept NATO forward strategy as desirable. They take into account the weaknesses of NATO's present forces. But they have focused on a more fundamental issue—how NATO forces are organized for forward defense, rather than how many forces there are. Anders Boserup, Professor of

Sociology at the University of Copenhagen and co-convener of the Pugwash group on conventional forces in Europe, is, together with the Norwegian Johan Galtung, a major representative theoretician of this more fundamental critical approach. Boserup attacks the basic concept of NATO defense—that peace and stability are promoted by a balance of forces between NATO and the Warsaw Pact. He argues that efforts to achieve a force balance always lead to instability rather than stability when the character and organization of forces on both sides is of the same type, especially when their relative strength is fairly close, Boserup contends that efforts to achieve symmetry of forces or a force balance create a high risk of pre-emption in a crisis situation and pre-emptive escalation once conflict begins, and that they also entail an upward ratchet effect in the arms race. In his view, stability and security depend not on a balance of similar forces, but on inequality between defense and offence, on the superiority of one side's defensive forces to the offensive forces of the other. When each side has offensive capabilities well in excess of the defensive capabilities of the other, a situation is created where there is a premium on taking the initiative in crisis situations, for example with air or missile attacks on the air defenses, airfields, missile sites and command facilities of the other side.

Boserup concedes that deterrence through a balance of power made some sense in pre-nuclear Europe. Today, he argues, deterrence through a nuclear balance of terror and the threat of retaliation still has relevance for the superpowers, but not for Europe. Present NATO and Pact conventional forces are of similar character, both emphasizing mobile armored formations used for Blitzkrieg offensive. Such forces are stronger in the offensive than the defensive, creating a situation of enduring instability. Arms control negotiations, with their typical fixation on numbers and on the effort to bring similarly organized forces into better balance, are not the answer. Even their success perpetuates the problem.[2] Horst Afheldt, a political scientist who has turned to defense issues, is another leading theoretician of alternative defense. As Afheldt puts it, a security policy which has the goal of building up all the military options at the disposal of the adversary leads to an arms race.[3]

In sum, the proponents of alternative defense argue that efforts to achieve a balance of similarly organized and composed forces inevitably lead to crisis instability and arms race instability.

The alternative defense answer to this problem is radical unilateral change in the composition of NATO forces, with emphasis placed on defense, to break the cycle of competitive force improvements and begin a build-down of the East–West military confrontation. Proponents of alternative defense believe that although individual weapons systems can obviously be used for offensive or defensive

---

[2] Anders Boserup, 'Non-nuclear defcnce', in Mary Kaldor and Dan Smith, eds., *Disarming Europe* (London: Merlin Press, 1982). See also A. M. Saperstein, 'Technology and organization in defense of Europe', in *Bulletin of Peace Proposals,* Vol. 16, No. 2, 1985.

[3] Horst Afheldt, 'Defensive Verteidigung als politisch-strategische Konzeption', in Hartmut Bühl, ed., *Strategie-Diskussion* (Herford-Bonn: Mittler, 1987).

purposes, it is possible to distinguish between offensive force configurations—emphasizing mobility and range—and defensive force configurations—emphasizing firepower, with limited mobility and range. They argue that a manifestly defensive system unsuited to attack will not be perceived as a threat by the Warsaw Pact, will reduce the risk of conflict, and will provide more effective resistance against possible attack.

5.3 Alternative Defense Analyzed

This article will concentrate on alternative defense proposals as they are relevant to the defense of the Federal Republic of Germany. Located in the area of the greatest East–West force concentration, the FRG—which has generated the largest number of specific proposals for alternative defense—is also the area where alternative defense proposals are put to their greatest test: the presence of large Warsaw Pact armored formations in forward positions in the German Democratic Republic and Czechoslovakia.

In the present treatment, alternative defense approaches will be divided into three main categories: (1) organized non-violent defense or 'social resistance', accompanied by total or partial elimination of present German and other NATO armed forces; (2) replacement of present Bundeswehr and allied armored formations by forces organized in a radically different way; and (3) modifications of present Bundeswehr and NATO organization, retaining a traditional armored component (though generally a considerably smaller one) but adding a forward shield of high-technology defenses. In each case, I will give a description of the general characteristics and objectives of the proposals, present some specific examples of each group, and evaluate them. I will then briefly describe the Soviet and Warsaw Pact reaction to this flow of proposals for alternative defense, look at the relevance of the proposals to arms control, and conclude with an overall evaluation of the possible future development and application of alternative defense.

Inevitably, given its origin in groups opposed to the decision of the West German and other NATO governments to deploy new American INF missiles in Europe, the entire subject of alternative defense has been politicized from the outset. Most statements on both sides have been so politically charged that it is difficult to obtain an objective reading of the advantages and disadvantages of specific suggestions. Moreover, any suggestion for far-reaching change in a NATO defense posture which has been operating for a period of over 30 years during which there has been no war in Europe has a heavy burden placed on it to prove its case.

The present alternative defense ideas have many forerunners. For social and political reasons similar to those of the framers of the American constitution, the European left has a long tradition of antipathy to large standing, heavily equipped armies headed by professional officers. In the closing years of the 19th century

August Bebel, founder of the German Social Democratic Party, argued for a militia system instead of standing forces. Before the First World War Jean Jaures, the French Socialist leader, argued for a purely defensive military structure, a militia composed in large part of reserve personnel, with those in frontier regions keeping their arms at home.

Over the years, NATO military experts too have proposed radical changes in the organization of NATO forces. In the 1950s, Marshall of the RAF Sir John Slessor proposed a highly trained home guard of limited mobility for Germany. In the 1960s, B. H. Liddell Hart proposed a citizen militia for Germany on the Swiss model, manning a deep network of forward defense posts. In the early 1970s, Kenneth Hunt of the International Institute for Strategic Studies proposed a three-layer organization for NATO's defense of Germany, to be composed of light formations in a forward belt along the dividing line, part active-duty units and part militia, a second belt composed of armored units, and a third rear area belt of reserve units. These ideas bear a marked similarity to the alternative proposals of a decade later.[4]

Proposals by German experts to restructure the defense of the FRG also have long-established antecedents. As the treaty negotiations to permit establishment of new German armed forces were being conducted between Germany and the Western occupying powers in the early 1950s, Colonel Bogislaw von Bonin, chief military planner of the Amt Blank, the precursor of the German Defense Ministry, proposed that Germany's contribution to NATO defense should take the form of an 80-km defensive belt emphasizing anti-tank weapons, mine-fields and defensive positions manned by light infantry and local militia.[5] Bonin was dismissed and his ideas never carried out. The reason is still relevant today to ideas on the reorganization of German defense—German forces organized as Bonin had proposed could not interlock with heavily armored American forces organized on Second World War lines. Chancellor Konrad Adenauer and his government considered it essential that the new German forces be organized so they could mesh with American and other allied forces in a joint defense concept which would fully involve the allies in the defense of Germany.

The nomenclature and vocabulary of the numerous proposals for restructuring NATO forces is wide-ranging and not standardized, but its very variety conveys an impression of their objectives. The descriptive labels used include: Defensive Defense; Non-Offensive Defense; Reactive Defense; Non-Provocative Defense; Structural Incapacity for Attack; Area Defense; Static Defense; Techno-Defense; Territorial Defense; Militia Defense; Non-Nuclear Defense; Autonomous Defense; and Transarmament. The use in this article of the term

---

[4] Details on the views of Bebel and Jaures are found on pp. 1–20 of Adam Roberts, *Nations in arms* (2nd edn, New York: St Martin's, 1986); early proposals for NATO reform are described on p. 271.

[5] The 'Bonin Plan' is briefly described in David Gates, 'Area defence concepts: the West German debate', in *Survival,* July-Aug. 1987. The article provides an excellent short survey of the alternative defence issue.

'alternative defense' as a collective term for this entire spectrum of ideas is legitimate, since one characteristic shared by all of these often widely divergent approaches is the desire to provide an alternative, not to forward defense—because all proponents wish in their own way to defend the totality of German territory—but to the present organization of NATO forces.

5.4 The Principles of Alternative Defense

The spectrum of alternative defense proposals is a wide one. Given such a range of approaches, what purposes or objectives do the proposals have in common?

One common characteristic of all the alternative defense proposals, understandable given the origin of most in the anti-nuclear movement, is that they wish to eliminate or at least radically reform the present NATO strategy of flexible response and possible first use of nuclear weapons. Proponents of alternative defense would replace flexible response with a no-first-use policy, and most would eliminate all remaining nuclear weapons from Germany. However, many versions envisage a sea-based American nuclear deterrent dedicated to European defense, with back-up nuclear deterrence through a minimum strategic force for the United States and the Soviet Union. To a large extent, alternative defense is also based on the related idea that by organizing their defense in a different way individual West European countries can be more self-sufficient in defense and rely correspondingly less on an alliance with a dominating American nuclear superpower. Underlying this national-minded objective is a change in attitude to the Soviet threat: the belief that Soviet attack on Western Europe for the purpose of conquest and holding the terrain to exploit its productive resources is highly unlikely, and that the more likely cause of war in Europe is a gradually emerging crisis, or the effects of a war outside Europe. Consequently, advocates argue, what is now needed in Europe is enhanced crisis stability and a reduction in the incentive to take a pre-emptive initiative. This objective can be achieved, first, through reorganization of NATO forces and, subsequently, through reorganization of the Warsaw Pact forces.

In most alternative defense proposals, restructured NATO forces are designed to operate in non-nuclear warfare, or at least to make Soviet use of nuclear weapons less probable by forgoing large and vulnerable force concentrations in favor of small, dispersed, concealed force groups. Most approaches urge that NATO or German armed forces should be limited in mobility and range, and not capable of deep counterattack on enemy territory. Given this, and the rejection of intermediate- and tactical-range nuclear armaments, the basic concept at the root of alternative defense seems to be dissuasion of attack through more effective denial of one's own territory, rather than the deterring of attack through the threat of retaliation. Most versions claim to be able to defend the German civilian population more effectively, some at lesser cost in weapons and personnel. Most rely heavily on high-tech microelectronic sensors and precision-guided munitions. Most are deliberately designed to further the arms control process, either indirectly by reducing presumed Warsaw Pact concerns and apprehension over the present NATO posture, or directly, by organizing NATO forces in a way that could be paralleled in the Pact forces. Most claim a higher degree of public acceptance and support, as well as higher morale among the defending forces.[6]

The following sections will deal consecutively with the three schools of alternative defense already briefly introduced above.

5.5 Group I: Non-violent Social Resistance

'Civilian-based defense', or 'social defense', based on training national elites in the techniques of non-violent resistance, is the most extreme of the approaches to the alternative defense of Europe. In this approach there are no armies, no weapons, and no allies, although friendly countries may wish to boycott possible Warsaw Pact invaders. In the FRG the leading proponent of these views, based on the concepts of the American sociologist Gene Sharp, is Theodor Ebert, professor of political science at the Free University of Berlin. Ebert argues that a leadership trained in advance in the techniques of non-violent resistance could effectively deny cooperation to an aggressor while maintaining the viability of its own society. By a deliberate policy of superficial collaboration without effective cooperation—'working to rule'—a population and its leaders can deny effective control over the production or use of a country, thereby denying it as effectively, and with less loss of life, as with military defense.[7]

Its advocates contend that this form of resistance can greatly increase the costs of military occupation, requiring even more control personnel. With trained leaders of social groups in the population—government officials, political party leaders, leaders of trade unions, churches and other social organizations—retaliation by terror would become increasingly expensive and ineffective. Ebert argues that non-violent resistance can also be combined with some forms of active defense.

This last idea has been developed by Wilhelm Nolte, a Bundeswehr officer serving at the Bundeswehr's Leadership Academy in Hamburg, in the concept of 'autonomous defense'. Nolte advocates a 30-km-wide network of static defenses along the inner-German and German-Czechoslovakian border, backed by a modest 'fire department' of mobile defense and guerrilla fighters. All cities of over 50,000 population would be declared open cities and would contain no military

[6] A useful and comprehensive p. 120 bibliography listing publications using most of these terms is Non-offensive defense bibliography, compiled by Björn Moller. It can be obtained from Björn Moller at the Centre of Peace and Conflict Research, University of Copenhagen, Vandkunsten 5, 1467 Copenhagen K, Denmark. Editor: This institute does not exist any longer.

[7] Gene Sharp, *Making Europe unconquerable* (Cambridge, MA: Ballinger, 1986); Theodore Ebert, 'Soziale Verteidigung als Politisch-Operative Konzept' in Bühl, ed., *Strategie-Diskussion*.

installations that could serve as targets. This point is a standard feature of most proposals for alternative defense. However, in Nolte's version urban populations, augmented by refugee populations from the countryside, would engage in passive resistance. Efforts would also be made to encourage passive resistance in rear areas of the Warsaw Pact.[8]

5.5.1 Non-violent Resistance Evaluated

To train leading FRG officials, trade union and church leaders in non-violent resistance techniques as a supplement for defense by the armed forces makes sense. But non-violent resistance has decisive shortcomings when it is advanced as the sole basis of defense.

(1) In the nature of a non-violent resistance program, the Warsaw Pact invader would not suffer retaliation. Even if unsuccessful, he could withdraw and try again without sustaining any real damage (2) The objective of a Warsaw Pact attack might be to pre-emptively deny the United States a foothold in Europe in a war motivated by the fear that conflict with the United States was unavoidable—not to conquer the FRG and other West European countries to exploit their productive capacity. Few of the alternative defense proposals, with their emphasis on denial of attacked territory rather than threat of retaliation, provide an adequate answer to such an attack. This is especially true of non-violent resistance, which would impose no barrier whatsoever to a movement of Pact forces to the English Channel and the North Sea.[9]

(3) A further objection, conclusive for most observers, is that non-violent resistance against forces of an authoritarian country like the Soviet Union, which has seldom shown itself squeamish over the use of violence, would make towering requirements on widespread personal courage and commitment, without the support of any military organizational framework to keep up morale. Such commitment is most improbable in large population groups. (It is not without significance that even the adherents of non-violent resistance in the German context often argue for a mixed system of military and non-violent resistance.) However regrettably, the conclusion is rapid and unavoidable that this approach is not a viable alternative to organized military defense, either as an effective recourse against armed invasion or one that could be acceptable to the German population.

---

8 Wilhelm Nolte, 'Autonome Abwehr', in Bühl, ed., *Strategie-Diskussion*.

9 This point is made by Michael Clarke as a more general criticism of alternative defence in an excellent review of the subject: Michael Clarke, *The alternative defence debate* (Occasional Paper No. 5, Armament & Disarmament Unit, University of Sussex, Aug. 1985), p. 9.

5.6 Group II: 'Classic' Models of Alternative Defense

The classic models of alternative defense substitute defense forces organized in a radically different way for the traditional heavy armored formations of the Bundeswehr and other NATO forces deployed on German territory. Armored units and attack aircraft are replaced by widely dispersed small units, either positioned statically or of limited mobility operating from protected concealed positions, backed up by a heavy deployment of electronic sensors and rocket artillery, using 'smart'[10] munitions designed for anti-tank and anti-aircraft use. Alternative defense concepts are anathema to supporters of NATO's recently adopted 'Follow-on Forces Attack' (FOFA) doctrine, which envisages long-range interdiction of Warsaw Pact reinforcements as they move forward. In turn, alternative defense advocates claim that FOFA increases the incentive for pre-emptive attack. In fact, both schools of thought are based on the same type of high-tech micro-electronic equipment, grouped under the heading of 'emerging technology' or ET.

Anders Boserup has provided a good summary of the conceptual basis of these proposals for a radical reorganization of NATO forces. Boserup[11] argues that West European forces should be organized to cope with a situation of sub-nuclear combat, in which nuclear escalation is possible but both superpowers wish to avoid it. Stability, he argues, is created when defense is stronger than offence on both sides rather than by matching offensive forces symmetrically to approach equality. Defense stronger than offence would not necessarily result from the type of weapons deployed, but from an organization of forces designed to fight defensively in a particular situation. These forces would need to be stronger in defense than the attacker forces, but at the same time so weak in offensive capability that they could not overcome the adversary's own defenses. Creating such forces, Boserup argues, would amount to imposing stability unilaterally.

This approach deliberately renounces retaliation as a deterrent. In Boserup's view, the objective of the armed forces in war is to produce a stalemate and to encourage political negotiation, avoiding escalation to nuclear weapons. The point is not to win victory—decisive engagements are avoided as risking escalation—but to create a stalemate pause in which politics can take over. This objective is embodied in a reactive strategy which relies on stationary or limited mobility force, to stalemate aggressive attack through attrition of the attacking force. The defending force is so organized that it cannot manoeuvre, concentrate or initiate a major battle.

The means of implementing this radically changed approach is area defense, deployed in depth. A large number of small units provide for area coverage. Firepower is provided by precision-guided munitions, and staying power and low

---

[10] Smart' munitions arc that category of weapon which home in on their targets independently of the aim of the person firing them.

[11] Anders Boserup, 'Non-offensive defense in Europe', in Derek Paul, ed., *Defending Europe: options for security* (Philadelphia, PA, London: Taylor & Francis, 1985).

vulnerability by concealment so the attacker cannot find targets. Firing platforms remain out of the battle area, dispersed and concealed,' sensors and forward infantry provide real-time information on enemy movements. The attacker has the choice of pushing through these defenses and taking increasing losses on his leading edge and flanks. Or he can expend a very large number of forces in a mopping-up operation. Or he can use 'carpet bombing'—impractical over so wide an area as long as conventional bombs are used. Boserup argues that, if defensive forces are so organized, there are no vulnerable installations and supply lines, and hence no need for aircraft to attack ground targets or assure air superiority, no airfields, no armored vehicles or warships. Precision-guided munitions, which no longer require indiscriminate mass fire, are the key. The attacker takes great losses. His large tanks, armored personnel carriers, trucks, aircraft and ships are seen first by the concealed defenders and are destroyed.

Horst Afheldt is the prototypical representative of this approach. In his 1976 book, *Verteidigung und Frieden* (Defence and peace), Afheldt developed a model of a network of autonomous '*Techno-Commandos*', small groups of soldiers scattered over the entire German territory with the objective of bringing the attack to a stop through attrition. There are no large units. Each of the approximately 10,000 *Techno-Commandoes* is composed of 20–30 soldiers. This force is equipped with a very large number of conventional rockets. Tactical nuclear weapons are eliminated. However, a second-strike capacity through sea-based US nuclear weapons would be possible.

In a 1983 book, *Defensive Verteidigung* Afheldt[12] revised his original concept and presented a 'transitional version'. In this, a network of *jaeger* or light infantry personnel organized in units or no more than 23 men are deployed along the German border. These units, equipped with light anti-tank and anti-aircraft weapons, with rocket tubes for mines and mortars, slow the forward movement of the attacking tanks and infantry. 20 km behind this initial network is a network of heavier rocket-tube artillery. Behind these, traditional armored units are deployed to cope with breakthroughs. Further groups of *Techno-Commandos* are dispersed over the entire German territory. Afheldt concedes that first use of nuclear weapons is possible as a political demonstration. Cities are declared non-combat zones, where the struggle is continued with non-violent means. Afheldt claims that even his transitional approach, which he calls a 'porcupine defense', poses far decreased demands on logistics. Reduced distances decrease the requirements for leadership and the need to move under fire. His *Techno-Commandos* conduct the battle from concealed, preconstructed shelters. There are no targets for the opponent's nuclear weapons or air force.

Afheldt is a theoretician; he has adapted his concepts to incorporate further elements of orthodox NATO force posture as he has attempted to meet criticism. For the true, unflinchingly consistent advocate of the high-tech defense we have to

[12] Horst Afheldt, *Defensive Verteidigung* (Reinbek, Hamburg: Rowohlt Taschenbuch Verlag, 1983).

turn to Norbert Hannig, a former Air Force colonel and engineer. For his 'defense wall', Colonel Hannig proposes the use of prefabricated sectional steel bunkers, mounted in a few hours by crew and capable of protecting their crew during observation, aiming and firing of their main weapons—rocket tubes, with interchangeable conventional warheads in heavy (180-km range), medium (40-km) and light (18-km) types. These replace both aerial bombs and artillery. These weapons can easily be re-aimed to give fire support to each other's areas. Some of these weapons would be mounted on rapidly elevated 'cherry picker' weapon-platforms (somewhat like those used to repair telephone and power lines), which would conceal and protect them up to the point of firing. Others would be mounted on heavy wheeled vehicles carrying their own excavating power shovels and capable of digging themselves fully into the ground in 2–3 h. By using this kind of linear defense at the borders, Hannig claims, NATO could dispense with tanks and fighter-bombers. A limited number of air superiority fighters would be retained to deal with high-level bombers.

Hannig has moved to meet one major objection to Afheldt's proposals; Hannig's defensive system is manned by the NATO allies with forces on German territory—Belgium, the Netherlands, the United States and the United Kingdom—*together* with the Bundeswehr. Hannig believes that the total number of fully mobilized NATO forces stationed in Germany organized in this way would not need to exceed 800,000 men, with 300,000 or less as active-duty soldiers in peacetime. (The present figure of NATO standing forces in the FRG and the Benelux countries is about 1 million.) Hannig calculates that the costs of his plan, including weapons systems, moveable shelters and ammunition, would total about $50 billion spread over a 15-year period. He suggests the costs could be shared by those NATO countries with forces in Germany.[13]

5.6.1 Classic Models Evaluated

Afheldt and Hannig add to Boserup's concept of a force primarily organized for defense the idea that technological progress, especially in the field of microelectronics and data-processing, makes possible specialized weapons for defense. The elimination of tanks and fighter-bombers removes dual-purpose weapons which could be used offensively as well as defensively. The approach meets the requirement of alternative defense that successful resistance must not be dependent on nuclear weapons, nor present targets for Pact nuclear weapons. Mobility of firepower substitutes for mobility of troops. The system is well insulated against the effects of conflict outside Europe, cuts pre-emptive pressures, and reduces the

[13] Norbert Hannig, *Verteidigen ohne zu bedrohen*, AFES, Institut für Politikwissenschaft, Universität Stuttgart, Nov. 1986.

need for massive mobilization in crisis periods and for reinforcements from the United States early at the outset of conflict.

Yet both concepts, especially Afheldt's, have been vigorously criticized. Afheldt's original concept was a complete break with traditional concepts of military organization, especially his idea of very small, autonomous *Techno-Kommandoes* set up in backyards and gardens throughout the FRG, waiting in relative isolation for an attack in their five-square-kilometer area of responsibility. Moreover, Afheldt's concept is essentially one of national defense, to be conducted by Germany on its own. It has no role for a supporting alliance, except in one respect. Afheldt's approach looked as though he were wholly dispensing with nuclear deterrence. But this was not the case; he merely pushed the deterrent to sea, by eliminating land-based weapons.

Among more specific criticisms, the techno-defense concept is said to rely too heavily on fixed assets, bringing with it the danger that Pact forces could drive a corridor through techno-defense by using chemical or nuclear weapons (although both Afheldt and Hannig provide a role for nuclear weapons as a deterrent against their use by the Pact) or a high-altitude air attack above the range of fixed ground anti-aircraft systems, which could also bomb population centers, leaving defended areas to be mopped up subsequently with no mobile force to counterattack against enemy breakthrough (a point met in the later version of Afheldt's approach, and also by the proposed capacity of the rocket artillery to shift targets rapidly).

A major criticism of the Hannig/Afheldt approach is that it relies too heavily on new electronic weapons and precision-guided munitions, many of which will not be readily available in fully developed versions for decades—a serious shortcoming for a concept so dependent on high-tech equipment. (Hannig argues that the equipment will be readily available within 10 years.) Some critics insist that precision-guided munitions may never fulfil their promise and may be insufficiently robust to meet battlefield conditions, night, bad weather, electronic countermeasures or even simple decoys. Adherents of NATO's present force posture consider that relying so heavily on a relatively static defense through high-tech weapons would fix all NATO's defense assets in one mode and deprive the defense of the flexibility of a mobile armored force. They argue that NATO's present mobile forces would be in a better position to fight on once their command centers were destroyed than if the command centers of Hannig's defensive wall were destroyed. (Hannig claims that his small dispersed command centers would present less of a target than present command installations.) Critics charge that the rocket weapons favored by Afheldt and Hannig would be less effective against enemy armor than tank-mounted cannons firing uranium slugs. Hannig counters that his armaments can hit tanks from underneath (through scatter mines) and on top, where they are less armored. (This dispute is a more specialized version of the current debate about the relative effectiveness of various anti-tank weapons against ceramic and explosive tank armor.) The techno concept is also faulted for logistic weakness, although it depends on dispersed pre-stored ammunition stocks.

There are more general criticisms of the techno-defense concept. Many critics consider that eliminating the possibility of counterattack on enemy territory would eliminate an important NATO deterrent and a capacity to push back an invading force out of German territory. A related view holds that it is difficult to keep up morale among military personnel when they are deployed exclusively in defensive positions, with no opportunity to retaliate against an aggressor. In response, Afheldt and Hannig claim that force morale would be higher in a situation less dependent on early use of nuclear weapons on German soil. Since nuclear weapons play a back-up role in both Afheldt and Hannig concepts, critics question whether an American president would be as likely to run the risk of using nuclear weapons in support of a defense which did not expose large American formations to being overwhelmed by Pact attack. Moreover, the original Afheldt concept was seen as edging towards neutrality and increasing the possibility of some form of creeping reunification of Germany, especially if, in the course of time, the Soviet Union permitted East Germany to adopt a similar posture.[14]

5.7 Group III: Modifications of Present NATO Force Posture

A third group of alternative defense proposals retains armored forces similar to those in the present formations of the Bundeswehr and other NATO allies—though in most cases considerably smaller ones—but interposes a barrier of static high-tech or light infantry defenses along the dividing line between the two forces to take the first impact of a Pact attack. For the most part, these proposals also limit nuclear weapons dedicated to European defense to sea-based weapons. In most cases the proposals represent efforts to meld components or elements of alternative defense, including high-tech weapons, with present NATO or Bundeswehr force organization.

The stated objective of most of these 'mixed' approaches is to improve conventional defense, to reduce reliance on nuclear weapons, improve crisis stability, increase political support, reduce costs, and introduce a degree of defensive specialization among the NATO countries, instead of concentrating on an all-purpose armored defensive/offensive force. Advocates of this mixed approach would do away with NATO's FOFA concept because, they argue, long-range interdiction bring with it a developed capacity to initiate an offensive. Most versions place far greater reliance on reserve forces, including local reserves to man border defenses, and also foresee a much diminished role for the NATO allies in ground defense.

The debate on alternative defense currently focuses on this group of proposals, because they make a stronger claim to replace current NATO posture than the

[14] Useful summaries of criticisms of alternative defence are contained in Gates, 'Area defence concepts'; in Christian Krause, *Strukturelle Nichtangriffs-Fähigkeit im Rahmen Europäischer Entspannungs-Politik* (Bonn: Studiengruppe Sicherheit und Abrüstung, Friedrich-Ebert-Stiitung, 1987); and in Günther Schmid, 'Die Bewertung alternativer Konzepte', in Bühl, ed., *Strategie-Diskussion.*

more radical alternatives of Afheldt (original version) and Hannig. In this section I will consider four of these 'mixed' alternative defense concepts—those put forward by the Study Group on Alternative Defense Policy, by Jochen Löser, by Albrecht von Müller, and by Andreas von Bülow.

5.7.1 The Study Group on Alternative Security Policy

Lutz Unterseher and the Study Group on Alternative Security Policy (SAS), a group based in the FRG and close to the Social Democrats, have put forward proposals for a defensive system of three belts: (i) a 30-km containment net (*Fangnetz*) of static area defense, manned by decentralized infantry detachments, partly active duty, partly reserves; (ii) a 50-km zone of rapid commitment forces (the 'fire department')—mechanized forces with a limited degree of operational mobility, including anti-tank infantry, tanks and helicopters, using obstacles, prepared shelters and heavy firepower rockets; and (iii) a rear area force designed to protect against sabotage and airborne landings. The scheme features decentralized ammunition and fuel depots and small units, with no fighter-bombers but light interceptor aircraft and ground-to-air missiles, and a reduced role for allied infantry in Germany, whose number would be reduced to 100,000 in peacetime. In addition to this force, the allies would be expected to specialize in aircraft and naval forces. Total Bundeswehr mobilization strength in wartime would be cut from 1.3 million to 680,000.[15]

5.7.2 Jochen Löser

Jochen Löser, a former Bundeswehr division commander, urges a more effective defense with greater emphasis on European contributions within the NATO framework. (Separately, he has also proposed the neutrality of central Europe as a political solution for the East–West confrontation.) Löser argues for a broad area defense of Europe with much increased conventional forces, complemented by a build-up of civil defense. This scheme divides the FRG into the three zones usual for this mixed approach: (i) a 'shield' of border area defense provided by active-duty light infantry Bundeswehr forces under NATO command, consisting mainly of local reserves with heavy use of obstacles; (ii) a zone of area defense with 'sword' forces composed of counterattacking armored brigades; and (iii) militia-type home defense units providing rear area security. Loser's scheme greatly increases the

---

[15] Studiengruppe Alternative Sicherheits-Politik, *Strukturwandel der Verteidigung* (Opladen: Westdeutscher Verlag, 1984); Lutz Unterseher, *Defending Europe: toward a stable deterrent* (Bonn: Studiengruppe Alternative Sicherheits-Politik, 1986).

strength of the Bundeswehr by doubling the present 36 brigades of the field army through creating an equal number of combat brigades for reserve personnel (which would require a large additional number of professional officers and non-combatants, as well as doubling the equipment holdings of the Bundeswehr, at considerable cost).[16]

5.7.3 Von Müller and the 'Integrated Defense Model'

Albrecht von Müller, of the Max Planck Institute, Starnberg, and the University of Munich, has made a consistent and determined effort to mesh alternative defense concepts with current defense strategy and organization. His approach, 'the integrated defense model', is organized into four zones: (i) a 5-km fire-belt at the border, with fire power from artillery rockets (multiple-launch rocket systems), attack drones, and mines of all kinds; (ii) a 'hook' or 'net' zone 25–50 km deep, manned by light infantry *Techno-Commandos* equipped with third-generation precision-guided munitions, mortars and scatter mines; (iii) a 60-km deep 'man oeuvre zone' with heavy armored units to cope with possible breakthrough and to expel a weakened invader from German territory; and (iv) a rear defense zone consisting of a network of mobile defense units to deal with sabotage and air landings and provide logistic support. The air force specializes in maintaining air superiority over FRG territory; there are no fighter-bombers. Müller considers his approach an advanced form of flexible response. However, like the other authors, he foresees a policy of no first use and sea-based deployment of nuclear weapons only as a deterrent against their use. He is silent on the role of NATO forces in his scheme.[17]

5.7.4 Andreas von Bülow

Andreas von Bülow, Bundestag member and defense expert of the opposition German Social Democratic Party (SPD), is the last of the integrative alternative defense thinkers to be presented here. As a Bundestag deputy and well-known

---

[16] Jochen Löser, 'Gedanken zur Weiterentwicklung von Strategie, Operation und Tacrik', *Allgemeine Schweizerische Militärzeitschrift,* Vol. 152, No. 9, Sept. 1986; 'The security policy options for non-communist Europe', *Armada,* No. 46; Jochen Löser and Ulrike Schilling, *Neutralität für Mitteleuropa: das ende der Blocke* (Munich: Bertelsmann, 1984).

[17] Albrecht A. C. von Müller, *The integrated forward defence* (Starnberg, 1985);'Confidence building hardware measures[1], paper for the 34th Pugwash Conference on Science and World Affairs, July 1984; 'Structural stability at the Central Front', Paper No. 13, Niels Bohr Centennial, University of Copenhagen, Sept. 1985.

member of the SPD opposition, von Bülow comes closer to the German government position on defense than other proponents of alternative defense, but there are still highly significant differences of approach. Von Bülow's concept does not have the explicit support of the SPD, although it has approved many of its component elements.

Von Bülow considers it essential for Germany to maintain the forward defense strategy and the full participation of NATO, with continued deployment of allied troops on German territory. But he terms the present forward defense position a Maginot Line composed of armored units, vulnerable to breakthrough because of the need to spread a limited number of units along the line of defense, with few operational reserves. Bülow argues that German forces should be more specifically configured to the defense mission. He points out that demographic pressures will decrease the number of Bundeswehr personnel on active duty. It will be necessary to make greater use of reserves. But only those changes in the Bundeswehr should be made which can be integrated into the NATO common defense. NATO, he thinks, should reduce the need for first use of nuclear weapons by making conventional forces stronger and withdrawing tactical nuclear weapons from Germany.

Von Bülow argues for the 'shield and sword' concept in conventional forces. Behind a 40–60-km shield of home defense regiments, mainly reserve forces, German and allied armored brigades provide the sword. The rear area is covered by a network of home defense forces into which combat support, medical and logistic troops are integrated. In the Bundeswehr, for reasons of rationalization and economy, the divisional level of command and organization is eliminated, as is the present organizational distinction between Germany's Field Army (under NATO command) and the territorial army (not under NATO command). Military district commands are also eliminated.

In von Bülow's proposals, the border-protecting shield forces consist of 200 German home defense regiments, each consisting of 2,000 men organized into eight motorized infantry companies. Each company covers an area 5 km wide and 10 km deep, is armed with fiber-optic-guided missiles, and is supported by combat engineers and heavy artillery weapons deployed behind, with armored brigades to the rear. In some sectors these armored forces are provided by NATO allies. The function of the home defense units is to reconnoitre and locate the enemy, direct artillery barrages, pin down and wear down the attack force, using cover and prepared positions. The home defense regiments consist of active-duty conscripts and reserves drawn from the local population intimately familiar with the terrain.

After extensive analysis of Warsaw Pact forces, von Bülow has concluded that the West would in fact have longer warning time of Pact attack than is generally believed. This increase in warning time would permit a decrease in the number of German active-duty personnel and an increase in the use of reserves to be called up in case of need. Standing forces are reduced by 100,000. Some armored units are converted to reserve status by having their active-duty personnel reduced, and the mechanized infantry (*Panzer Grenadiere*) currently used in the Bundeswehr to accompany forward-moving tanks are converted to light infantry for border protection.

A major objective of von Bülow's approach is to ensure expansion of the Bundeswehr to its full mobilizable strength of 1.3 million through an expanded reserve structure. Conscripts are trained for the same specialized function they will have in the reserves, like manning anti-tank weapons and placing mine barriers, and for this they would be mobilized (not now done). Conscript personnel in active-duty units will move into sister reserve units after an active-duty training period cut by half to 7 months from the present fifteen (which, currently, is soon to be increased to eighteen to deal with manpower shortages by keeping conscript personnel in service longer). Von Bülow argues that the best way to deal with the Bundeswehr manpower shortage is by cutting the number of active-duty service personnel and increasing the period over which former conscripts would be liable for reserve service, as well as increasing reserve training exercises.

Von Bülow wants to keep the allies involved in the forward defense of the German border. With this in mind, he would place the border defense network of home defense regiments under allied commanders from NATO defense sectors. Von Bülow considers that the establishment of a defensive zone in each alliance's territory would facilitate deep negotiated reductions in the armored 'sword' forces on both sides—an explicit objective also shared by Unterseher and his SAS colleagues and by Albrecht von Müller. In the long term, von Bülow wants to eliminate the perpetuation of Europe's division into two heavily armed social systems 'in the grip of the two superpowers'. He hopes that both Soviet and American forces will have disappeared from central Europe by the end of the century.[18]

5.7.5 Group III Proposals Evaluated

An intentional feature of most of the approaches described in this section is that they are variations on existing NATO defense organization. Unlike those of preceding groups, these are not proposals to replace that organization with something radically different. All depend to some extent on nuclear deterrence, and all envisage some use of traditional armored forces. Indeed one criticism, applicable especially to the variant proposed by General Löser, who would retain all the

[18] Andreas von Bülow, 'Skizzen einer Bundeswehr—Struktur der 90er Jahre' (paper, Bonn, Nov. 1984); 'The SPD's views of the defense of NATO's Central Region', paper prepared for presentation to the Institute for Defense Analysis, Washington, DC Mar. 1987; 'Defensive entanglement: an alternative strategy for NATO', in Andrew Pierre, ed.. *The conventional defense of Europe* (New York: Council on Foreign Relations, 1986).

Bundeswehr armored units, is that to the extent that such proposals would supplement existing NATO armored forces by adding an effective new shield force on the German border, they would, by the same token, improve NATO prospects for an offensive or deep counterattack into Warsaw Pact territory.

Despite the compromise they represent, this group of proposals has attracted more criticism than the extreme alternatives—partly because they are eclectic and appear more plausible. Partly this is because the proposals involve domestic party politics: most of their proponents are identified more or less closely with the SPD German opposition.

The Social Democrats endorsed the general concept of alternative defense in their 1986 security program, which argues that the armed forces of NATO and the Warsaw Pact should be so organized, structured, trained and equipped as to unmistakably serve defensive purposes and be incapable of mounting a large-scale attack. The SPD program calls, as does von Bülow, for the removal of intermediate-range and battlefield nuclear weapons from the FRG, a reduction in the length of conscript service, and a reduction in the Bundeswehr's active-duty personnel in the interests of developing a stronger reserve capability.

It is relevant to the question of political support for alternative defense in Germany that the German Free Democrats (FDP), coalition partners in the present government, have also endorsed alternative defense in general terms. President Richard von Weizsäcker, much respected despite the limited power of his office, also gave a general endorsement to the concept of non-offensive defense when he welcomed East German leader Erich Honecker to Bonn in September 1987. Later the same month in the United States, Chairman Les Aspin of the US House Armed Services Committee, in a speech to a colloquium organized by the American Association for the Advancement of Science, urged a framework for the reorganization of NATO defense which resembles some of the alternative proposals of Group III reviewed here.

There are clear military drawbacks to this group of alternative defense proposals. Most of them make heavy cuts in the number of active-duty soldiers in the Bundeswehr, to an extent which may well go beyond the requirements imposed by the population crisis and the squeeze in the conscript age group. They propose spending less, rather than more, on conventional defense, and would make defense still more dependent than at present on the mobilization of reserves, something that may be difficult to decide at times of crisis. By cutting back on the size of NATO armored forces and, in some cases, the mechanized infantry units that accompany them, these proposals would deliberately strip NATO forces of their capacity for counterattack, a capacity which presumably constitutes a deterrent to Pact attack. (Advocates of alternative defense consider this cutback a positive contribution to stability.)

The alternative defense proposals described here assume that land-based nuclear weapons will be eliminated from German territory. In the past, opponents of alternative defense have criticized this aspect as 'decoupling' and seriously undercutting the American nuclear guarantee. But if the US-Soviet agreement to eliminate INF and SRINF missiles, currently under negotiation, is finalized, the tactical nuclear weapons currently deployed in Germany will remain in politically prominent isolation. (Aircraft delivery systems will also remain.) Tactical nuclear weapons are already highly unpopular in Germany because of their short range and probable effects on the general population. Indeed the German government, pushed by conservative opinion in its own ranks, has adopted the Social Democrats' argument that the FRG should not be singled out as the sole NATO country which can be targeted by tactical nuclear weapons. Consequently, in disagreement with most other NATO countries, it has announced its determination to negotiate reductions in these weapons with the Warsaw Pact, which has itself proposed reductions in the forthcoming NATO-Pact Atlantic-to-the-Urals force reduction negotiations. It is fair to forecast that sooner or later the tactical nuclear weapons deployed in Germany will be reduced or even eliminated. When that happens, a major source of German opposition to alternative defense will have disappeared.

Another criticism of this group of alternative defense proposals is that they may provide weaker initial defense against Pact conventional attack than NATO's present mobile forces, through their policy of deploying reservists along the border in the area of first impact where they may not perform well. Critics comment that light infantry troops are highly trained elite forces. It is one thing if such elite troops are used in the shield function, and another if the shield function is to be provided by reservists. Other critics argue that, with so many NATO assets in fixed defenses against tank warfare, the Pact could change its tactics and go in for airborne assault or simply send the infantry ahead of the tanks. (Alternative defense advocates claim their proposals would cope with both possibilities.) Critics argue that fixed installations and zones of static defense combined with a smaller retaliatory force would make it easier for an aggressor to calculate the costs of attack, and thus weaken deterrence in this specific way also. (This criticism would not apply to versions, like that of General Löser, where the shield forces supplement existing armored forces.)

Critics also point out that chemical weapons or tactical nuclear weapons could be used to create a corridor through the fixed targets provided by shield forces, whether static or of limited mobility. Yet the critics themselves are now concerned about possible Soviet use of chemical, nuclear or conventional warheads against other fixed targets—anti-aircraft munitions airfields, ammunition depots in rear areas in the FRG. Most alternative proposals do away with these targets, and it is questionable which approach—mobile armored units or dug-in infantry and firing platforms protected by prefabricated shelters—would have a lower casualty rate in the face of concentrated artillery fire or air bombardment.

It is relevant to our assessment of this integrative or eclectic group of alternative defense proposals that many of their individual tactical (but not strategic) elements have been proposed by more orthodox critics of NATO's present force posture. Bundeswehr General Uhle-Wettler and Steven Canby have proposed the use of light infantry in border areas.[19] Canby and David Greenwood have proposed organizing Bundeswehr reserve manpower to form more mobilizable combat units and greater force specialization among the allies.[20] Leonard Sullivan has energetically advocated a high-tech barrier of sensors and rocket artillery as a shield for a back-up force of mobile armor.[21] John Keegan and Trevor Dupuy have argued for a belt of light field fortifications and obstacles along the NATO side of the dividing line—equivalent to the mine belts and dug-in shelters proposed by most advocates of alternative defense.[22]

A central problem with the ideas of the alternative defense group is money. Alternative defense adherents want to spend less on ground forces. In the end, it comes down to a question of resource allocation. The defense funds of the FRG, and of other NATO countries, are limited and likely to decrease. No defense planner wants to put the majority of his resources into new equipment never tested in combat, no matter how promising. That is the main problem with Colonel Hannig's well thought out proposals. Those who want to concentrate scarce resources on armoured forces as the most flexible and therefore most rational defence investment have the offensive and defensive experience of the Second World War and of warfare in the Middle East as justification. Yet, as Colonel Hannig and von Bülow point out, the use of mobile defence, which is expensive and consequently scarce, will create the very gaps in NATO defenses which can create openings for armored Blitzkrieg by the Warsaw Pact.

The present problem for NATO force planners is that, although they may see the theoretical benefit of shield assets, they do not now have enough money to field at an adequate level the basic armored force on which they rely; and they do not want to subtract from these limited funds to invest in limited mobility defenses of any kind.

Yet that consideration does not justify total concentration on armored forces. Expert after expert has urged greater emphasis by NATO forces on rapidly emplacable obstacles, scatter mines, pipe explosives, and on strengthening combat engineer components. These possibilities deserve more than the meagre resources now devoted to them, even at the cost of limited cuts in funds for armored forces.

---

[19] Steven L. Canby, 'Territorial defense in Central Europe', *Armed Forces and Society,* Fall 1980, Vol. No. 1; 'Military reform and the art of war'. *International Security Review,* Fall 1982, Vol. 7, No. 3; S. Canby and Ingemar Dörfer, 'More troops, fewer missiles', *Foreign Policy,* No. 55, Winter 1983/4.

[20] David Greenwood, 'Reshaping NATO's defences', *Defence Ministers and Chiefs of Staff* No. 5, 1984; 'Towards more specilization in NATO', *NATO's Sixteen Nations,* July 1986, Vol. 31.

[21] Leonard Sullivan, Jr., 'Allied burden-sharing: another view', *Atlantic Community Quarterly,* Spring 1986.

[22] John Keegan, 'Soviet blitzkrieg: who wins?', *Harpers,* May 1982; Trevor N. Dupuy, 'Strategy for victory or defeat?', *Air Force Magazine,* April 1983, Vol. 66, No. 4. See also John C. F. Tillson, 'The forward defense of Europe', *Military Review,* May 1981.

The criticism of alternative defense which perhaps is held most firmly is against the conscious, programmatic reduction of dual-use, offensive/defensive weapons like fighter-bombers and armored units in the interest of developing a non-provocative defense with the structural incapacity to attack. The intellectual argument for doing this is understandable, but most NATO military commanders would not be prepared unilaterally to give up or curtail their means of dealing with possible Soviet breakthrough, pushing an invading force back out of German territory, or retaliating against the attacking force in its own base area.

This fundamental criticism is intensified by a further major objection. Nearly all alternative defense programs posit a weaker and attenuated relationship between the FRG and its NATO allies. In particular they envisage Germany distancing itself from the United States. They either exclude the allies entirely from the defense process (like the original versions of the Afheldt plan), reduce the allied presence (as in the SAS approach), or announce their desire to dispense with the allies at some further point (as in the case of Löser and von Bülow). A reduction in the American troop presence in Germany may be a long-term possibility, but most Germans would not wish to encourage it, especially not unilaterally. The alternative defense approach also provides only a limited basis for intra-European defense cooperation, although some versions, like that of SAS, explicitly call for specialization by other European allies in air force and navy. On the other hand, it should be noted that the type of German forces foreseen by alternative defense adherents, with limited mobility, range and offensive capacity, could prove reassuring to Germany's neighbors in East and West.

NATO commanders might see alternative defense differently if it was also put into effect on the Warsaw Pact side. This point brings us to consideration of measures of force restructuring negotiated with the Warsaw Pact.

5.7.6 Arms Control Aspects

Depending on their scope and dimensions, parallel action by the Soviet Union and other Warsaw Pact states to restructure and redeploy their armed forces along alternative defense lines could change Western officials' predominantly critical reaction to alternative defense as a unilateral program. That would be the case, for example, if the Warsaw Pact proved willing to make drastic reductions in its armored forces, tactical missiles, self-propelled artillery, attack helicopters and attack aircraft deployed in central Europe and along borders with NATO countries—thus putting into practice the principle at the heart of alternative defense, that defensive forces should be stronger than offensive forces. If it could be assured that the mobile, forward-moving and terrain-gaining forces of each alliance were weaker than the static defense forces of the other, much higher crisis stability in conventional forces would result. A main objective of most of the German alternative defense schemes, especially those that envisage a buffer belt of static defenses or defenses with limited mobility backed up with armored forces, is that these models should be applied by both alliances, with each reducing its mobile armored force to a low level. What is the Soviet and Pact position on this issue?

In the 1980s, against the background of their earlier support for the Rapacki plan for the complete denuclearization of central Europe, the Soviet Union and its Warsaw Pact allies began to endorse in more explicit form the concept of exclusionary or 'thin-out' zones along the dividing line between the two alliances in central Europe. Such zones are relevant to alternative defense concepts because most of them posit one or more shield zones along the dividing line between the two alliances, where static defenses, but not heavy mobile armor, would be deployed. They are thus a possible way of contractualizing an agreement between the alliances to move into an alternative defense posture.

The Soviet Union endorsed the June 1985 draft treaty between the West German Social Democratic Party and the ruling East German Socialist Unity Party (SED) providing for the establishment of a chemical-weapon-free zone in central Europe. More importantly, it has endorsed the proposals of the 1982 Palme Commission and the October 1986 SPD-SED framework agreement for a 300-km wide zone in central Europe free of nuclear weapons—a concept subsequently formally advanced by the governments of Czechoslovakia and East Germany. The May 1987 communique of the Warsaw Treaty Political Consultative Committee urges:

> a nuclear-weapon-free corridor along the dividing line between the Warsaw Treaty and NATO countries that would be 300 km wide (150 km on either side), providing the withdrawal, on a reciprocal basis, of all nuclear weapons, i.e. nuclear munitions, including nuclear mines, shorter-range and tactical missiles, nuclear artillery, nuclear-armed tactical strike aircraft and nuclear-capable surface-to-air missile systems.

The same communique contains a broader, more general formulation which would appear to endorse the concept of zones from which most heavy conventional weapons as well as nuclear armaments would be excluded. The communique says,

> These talks should also cover a number of high priority measures designed to lower the level of military confrontation and avert the danger of surprise attack, to ensure the mutual withdrawal of the most dangerous offensive weapons from the zone of direct contact between the two military alliances and to reduce the concentration of armed forces and armaments in this zone to an agreed minimum level.

The Soviet Union has not only urged exclusionary zones and corridors in central Europe, it has moved to endorse the general concept underlying alternative defense proposals, the idea that NATO and Pact forces should be configured for defense and not for offence. This is a marked departure from earlier treatment of this subject by Soviet officials and academicians. Traditionally, both groups have expressed great skepticism about the value or practicality of the alternative defense approach, and have insisted that priority be given to negotiated reductions of armed forces. However, in his February 1987 speech to the Moscow 'International Forum', Mr. Gorbachev said that NATO and Pact armed forces should be reduced in a way which would exclude the possibility of surprise attack and remove the most dangerous offensive arms from the zone of contact. There should be a balance of reasonable sufficiency in constantly declining forces. He added: 'Quite naturally, military doctrines must be of a defensive nature.'

The theme of defensive force postures is more fully developed in the May 1987 communique of the Political Consultative Committee, which urges 'reduction of the armed forces and conventional armaments in Europe to a level where neither side maintaining its defense capacity *would have the means to stage a surprise attack against the other side or offensive operations in general*' (emphasis added). The communique ends with an invitation to NATO states to enter into consultation in order to compare the military doctrines of the two alliances as a means to reduce mutual suspicion, ensure a better perception of each other's intentions, 'and to guarantee that the military concepts and doctrines of the two military blocs and their members are based on defensive principles. The Polish Jaruzelski Plan, put forward by Warsaw in 1987 with the approval of its Pact allies, also endorses the concept of non-provocative defense in general terms, calling for strictly defensive military doctrines.

In discussions with Europeans and Americans in the spring of 1987, parallel to these official statements, several Soviet officials and academicians expressed explicit approval of and support for the concept of alternative defense. For example, in May 1987, at a conference near Hanover, Ambassador Oleg Grinievsky, former chief Soviet representative at the Stockholm Conference on Disarmament in Europe, said that East and West should give serious study to the concepts of non-aggressive defense in discussion of military doctrine between the pacts. There was a need, he said, for both sides to change their doctrines and give defense clear dominance. Victor Vassiliev, of the Soviet Committee for European Security and Cooperation, said that Western concepts of defensive defense represented a worthwhile search in a reasonable direction; Soviet analysts thought these concepts could form an important component of the 'new Soviet thinking'.[23] Many similar comments have been made in recent months to American and European researchers. It is notable that East European defense experts, especially in Czechoslovakia, Hungary and Poland, have shown increasing interest in this subject, and that an institute for its study is to be established in Prague.

Thus far, however, these positive statements have been cast in general, noncommittal terms. Under questioning, Soviet and Pact officials and academicians are unable to state with any specificity what kind of defensive force posture they

[23] Author's own notes of conference.

would envisage for NATO or, more significantly, for their own forces. They do say that practical application of 'offenceless defense' would mean large Soviet and Pact reductions in tanks, artillery and heavy mobile armaments. Soviet officials are likely to become somewhat more explicit in the coming months. At present it is not possible to discern whether the indications of growing Soviet interest in alternative defense merely articulate a posture designed to make the Soviet Union appear more cooperative to the European public interested in alternative defense, or whether they represent evidence, however inconclusive, of Soviet interest in building down the NATO-Pact confrontation.

In assessing the possibilities for future development of the Soviet position, it should be kept in mind that the Soviet armed forces have a confirmed attachment to tanks and defense through armored counterattack which is, if possible, even stronger than that of senior NATO officers. Moreover, discussion here has made it evident that a serious Soviet move towards some version of alternative defense would entail very large negotiated (or unilateral) reductions in Soviet heavy armaments and their destruction or storage. In the Atlantic-to-the-Urals force reduction talks, the Soviet Union will be in a position to make proposals showing what it is prepared to reduce. Yet at the time of writing it does not seem likely that it will at an early point indicate willingness to agree to the very large reductions in Pact armored equipment—five times or more as large as the equivalent NATO reductions—which would be needed to reach a position of parity with NATO. And NATO governments are unlikely to find the idea of military zones or belts where heavy mobile armaments are restricted attractive if they involve only a limited pullback of Pact forces without deep reductions in their armaments.

5.8 Conclusion

In their unilateral form, the more far-reaching proposals for alternative defense are not likely to be considered an adequate answer by political opinion in Germany or other NATO countries to Warsaw Treaty forces in their present size and configuration. As time goes on, individual aspects of these proposals—increased use of light infantry, increased attention to rapidly emplacable obstacles, sensors, precision-guided munitions and cheap rocket tubes—are likely to be incorporated piecemeal into existing NATO defenses. This process is already under way and may intensify as the population shortfall in Germany makes itself felt. In coming decades, step-by-step adoption will create more confidence in the new technology and the theoretical possibility of committing NATO increasingly to this defensive concept. However, strong resistance to complete and exclusive reliance on untried technology will continue; some armored forces are likely to stay in a mixed system.

What of the proposals' applicability to East–West negotiations? Even if alternative defense approaches were considered as something which could work in joint application by both alliances together, alternative defense is better suited to a situation in which the East–West military confrontation in Europe has *already* been built down to some extent than a role that is a means of achieving that build-down. In other words, alternative defense is more credible as the outcome of a decade of successful reduction negotiations, showing the way for the configuration of both alliance forces for maximum stability, than as a basis for achieving those reductions by negotiation. Theoretically, both alliances could agree on a few basic concepts of non-offensive defense and then seek to apply those principles in reducing and reconfiguring their forces. Tactically, it might be a good negotiating approach for one side or the other to advance this idea in the forthcoming force reduction negotiations. Yet it is improbable that these basic concepts are sufficiently acceptable to the military leadership of either alliance to serve as the main vehicle for their approach to the negotiations. This is certainly true of NATO, though Mr. Gorbachev has an affinity for futuristic concepts. What is more likely is that both alliances will make extensive use of the concepts of alternative defense as supporting arguments for more specific reduction approaches.

None the less, as the new Atlantic-to-the-Urals negotiations go through their early stages, the advocates of an alternative defense for Europe will be busy recasting their proposals into forms suitable for bilateral rather than unilateral adoption. Given their political support and the likelihood of continuing NATO-Pact confrontation in Europe at some level, these concepts are likely to have a long life. If, like their MBFR predecessors, the new East–West negotiations do not bring impressive results, and yet at the same time an INF agreement and other US-Soviet arms control agreements are concluded and implemented and there is no recurrence of major East–West tension, proposals for unilateral alternative defense will return in full vigor in Western Europe, with more public support than they now have. Given the practical focus of these plans on the defense of Germany, the attitude of the German public will be decisive. In the long run, if NATO forces shrink through economic attrition or the withdrawal of American forces outside the framework of negotiation, movement toward piecemeal adoption of these concepts, especially in versions like those of von Müller and von Bülow, will probably be accelerated.

In the meantime, those seriously interested in tracing the possible future course of the East–West confrontation owe more recognition to the various architects of alternative defense, most of whom have been attacked as dangerous visionaries, than they have so far paid. The proponents of alternative defense have made two major contributions to the debate on the defense of Europe. First, they have advanced what appears to be a telling criticism of balance-of-power theory, identifying one of the factors which has helped to maintain the military confrontation in Europe for a full decade after the political confrontation passed its peak with West Germany's conclusion of treaties with the Soviet Union and

Poland and the Berlin agreement. In this context they have advanced the most cogent exposition yet for the Soviet Union of what Westerners do not like about the Soviet military posture in Eastern Europe. Secondly, the alternative defense authors have given us some models to aim towards, some ways of visualizing how the East–West confrontation could look after a considerable build-down. These contributions are important.

Jonathan Dean testifying before the U.S. Congress as the Arms Control adviser of UCS, Permission to use this photo by courtesy of Union of Concerned Scientists in Washington, D.C., see at: http://blog.ucsusa.org/ambassador-jonathan-dean-390

Chapter 6
Berlin in a Divided Germany: An Evolving International Regime (1988)

In 1987 a city in Central Europe vigorously celebrated its 750th anniversary, despite the fact that that city, Berlin, no longer exists, at least in agreed East–West terminology.[1] In discussing East–West relations, Western officials do speak of 'Berlin,' while Soviet officials refer to 'Berlin West.' Western officials are referring to 'Greater Berlin,' the capital city of the Third Reich that was divided into four sectors of military occupation at the end of World War II. Soviet officials are referring only to the three Western sectors of Berlin. For them, there is only one Berlin—the capital of the German Democratic Republic (GDR), an area known to Western officials as the "Eastern Sector of Berlin."

These divergent designations reflect the division of Berlin and of Germany itself into two radically different political systems. They also reflect differences in basic legal definition of the Berlin situation. In the Western view, the rights in Berlin of the three Western allies—and of the Soviet Union—are original "rights of conquest" directly derived from their participation in the World II victory over Nazi Germany. In the Soviet view, the rights of France, the United Kingdom, and the United States in the three Western sectors of Berlin are not original rights but secondary rights, derived by agreement from the Soviet Union, which conquered and occupied Berlin at the end of the Second World War.

These East–West differences are and remain so fundamental that the formulators of the Quadripartite Agreement between France, the United Kingdom, the USSR, and the United States, signed in Berlin on September 3, 1971, could agree

[1] This text was first published as: "Berlin in a Divided Germany: An Evolving International Regime," in Alexander George, Philip Farley, and Alexander Dallin (Eds.): *U.S.—Soviet Security Cooperation—Achievements, Failures, Lessons* (New York—Oxford: Oxford University Press, 1988: 83–105. By permission of Oxford University Press, USA; see at: www.oup.com.

H. G. Brauch and T. Grimwood (eds.), *Jonathan Dean*,
SpringerBriefs on Pioneers in Science and Practice 19,
DOI: 10.1007/978-3-319-06662-2_6, © The Author(s) 2014

only on the term "relevant area" as the subject of the agreement on which they had been working for some 16 months.

Yet when officials of both East and West refer to Berlin's 'status,' they are referring to the same thing, the ensemble of rules and procedures that has grown up around divided Berlin. These rules and procedures, which prescribe and circumscribe the actions and interactions of representatives of France, the United Kingdom, the United States, the USSR, and of German officials in both parts of the divided city and country, constitute what has been called a security regime.' In the international context, a security regime can be seen as a set of rules that allows two or more countries to live more safely with a continuing and dangerous dispute—in this case the division of Berlin—which cannot be fully resolved to the satisfaction of all parties or dealt with by force except at unacceptable cost. It is a more explicit, more deliberate version of what is often called a "modus vivendi."[2] Divided Berlin exists and is governed under such an international regime.

Throughout the 40 years of the Berlin regime's existence, there have been repeated East–West differences over it. At least twice—in 1948 with the Berlin blockade, and in 1958–1961 when Nikita Khrushchev threatened to make unilateral changes in the status of Berlin—the regime threatened to collapse, and there were real possibilities of East–West conflict. In each case, the Soviet Union retreated from extreme positions and the Berlin regime was re-established, with modifications.

In fact, the Berlin regime has changed importantly during the past 40 years. Its original form was a set of agreed rules by which the four victorious wartime allies governed the former German capital. This arrangement lasted for 3 years, from June 1945, when the Western allies entered Berlin, to June 1948, when the Soviets blockaded the city. At that time, the Berlin regime underwent fundamental change. Instead of an arrangement for joint administration of a unified city, it became a four-power modus vivendi for administering a divided city in a divided nation. In 1971, the Quadripartite Agreement made a major adjustment that reduced both the risks of the existing regime and the hardships it caused for Berliners. It did so primarily by converting the existing Four-Power regime to a two-tier arrangement, inserting a second tier, a German one, associating the two German states that have emerged on the territory of divided Germany with the operation of the Berlin regime. In essence, though there are differences in the legal status of participants, the Four-Power Berlin regime was converted to a Six-Power regime. At the same time, both East and West accepted constraints on achievement of their ultimate aims with regard to Berlin, the USSR and GDR on their aim of incorporating the

[2] See "Security Regimes" by Robert Jervis, *International Security* 36, 2 (Spring 1982). I am indebted to Paul Viotti, an originator of the concept, for the description of the present situation in Berlin as a security regime. See his "Berlin and Conflict Management with the USSR," Orbis 28, 3 (Fall 1984). For useful bibliographies on postwar Berlin issues, see Jack M. Schick, The Berlin Crisis 1958–1962 (Philadelphia: University of Pennsylvania Press, 1971); and Honorc M. Catudal, A Balance Sheet of the Quadripartite Agreement on Berlin (Berlin: Berlin Verlag, 1978).

Western Sectors in the GDR through increasing GDR control over the access routes, and the Federal Republic by suspending its claim that Berlin was an integral part of the Federal Republic.

This chapter tells the always absorbing, often dramatic story of the Four-Power Berlin regime and of its conversion to a Six-Power regime, a development that can be described either as a succession of regimes or as an evolving single regime. It is the story of the deliberate, cooperative stabilization of a dangerous point of East–West confrontation. This development, although in many respects unique, could, like the SALT agreements, serve as a model for deliberate efforts to insulate specific areas of friction and competition, such as the Persian Gulf, from the daily flow of continuing U.S.-Soviet competition, and establish rules for administering these issues so as to reduce their potentiality for triggering a confrontational crisis. Here we seek to identify the elements that constituted obstacles to this effort and those that made success possible.

6.1 The First Berlin Regime

The original structure of the Berlin regime was established by the European Advisory Commission, which started its work in London in January 1944 as a joint commission of the United Kingdom, the United States, and the Soviet Union to coordinate postwar policy in Europe. On September 12, 1944, as World War II still raged in Europe, representatives of the three countries in the Commission signed the Protocol on Zones of Occupation and Administration of the Greater Berlin Area.[3] The Protocol established a unified administration, the Kommandatura, for 'Greater Berlin,' or 'Gross-Berlin,' the capital area as geographically delineated in prewar German law. It divided the city into American, British, and Soviet sectors of military occupation to be administered by the individual sector commanders, not autonomously, but under the instruction of the joint Kommandatura. In the Western view, at least, Berlin was to be a unified city under common rule, not associated with the occupation zones of the three powers or part of them. The three sectors into which the city was originally divided became four in July 1945, when France was associated with the administration of the city and given an occupation sector of its own.

There was no provision in the Protocol of September 12, 1944 for access to Berlin, either for movement of Western military forces in and out of the city from their respective zones of occupation, or for movement of German civilians or civilian goods. On June 14, 1945, before the planned withdrawal of American troops from the territory of the zone allocated to Soviet occupation, President

[3] The meaning of the term 'regime' as it is used here is well captured by the term used for the Quadripartite Agreement in the Federal Republic, where it is called the 'Berlin-Regelung'; a 'Regelung' is a set of rules for systematic treatment of some issue; the equivalent French word is 'reglement'.

Harry Truman sent a cable to Josef Stalin in which he asked for arrangements to provide unrestricted access to Berlin for American troops by air, rail, and road on agreed routes from Frankfurt and Bremen. Churchill sent a similar message regarding access for U.K. forces. Stalin's reply to Truman on June 16 suggested two weeks delay in the pullback of U.S. and U.K. troops to allow for clearance of mines from Berlin; he pledged that "all necessary measures" would be taken by Soviet authorities, but he did not specifically address Truman's proposal for agreement on free access.

On June 29, 1944, the U.S. and U.K. commanders, General Lucius Clay and Lt. General Sir Robert Weeks, met with Marshal Georgi Zhukov, the Soviet commander-in-chief, at Zhukov's headquarters in Berlin. Oral agreement was reached whereby the two Western forces would have one road, one rail, and one air route to Berlin. (They asked for more, but did not get it.) Zhukov said Soviet military personnel would check documents of British and American personnel, but there would be no inspection of cargo. These points were not committed to writing, and the oral agreement covered only allied military traffic—not interzonal movement of Germans, of which there was little at the time. Yet at the same time, the Western allies were required by circumstances—the nearly complete destruction of Berlin and of the German system for distribution of fuel and foodstuffs—to assume from the outset full responsibility for the supply of the civilian population of their sectors. These circumstances, the dependence of the Western sectors on the Allies, the absence of contractual provision for access, and the location of Berlin in the midst of the Soviet occupation zone, were the source of most of the dramatic developments surrounding the Berlin issue as the wartime allies fell into disagreement.

On November 30, 1945, the Allied Control Council approved in writing the establishment of three air corridors instead of the original single corridor, which had proven insufficient to cope with the traffic. Subsequently, agreement in writing was reached to establish the Berlin Air Safety Center (BASC) to coordinate control of all flights in the congested Berlin Control Zone, the airspace up to 10,000 feet in a circle 20 miles from the Control Council building. The BASC, despite many ups and downs in the smoothness of its operations, continues to function today as part of the Berlin regime.

The single rail line from Helmstedt to Berlin via Magdeburg, which had been agreed to in the Clay-Weeks-Zhukov conversation, also proved inadequate to the demands of Western traffic. Agreement was reached in writing on September 10, 1945 to permit three trains in each direction for Western military use, eight for coal used for both civilian and military purposes, and five for food for the civilian population, or a total of sixteen trains per day in each direction, or sixteen 'paths,' as they were called.

Road, rail, and waterway traffic of German civilians between the Western and Soviet zones was left to be regulated by occupation authorities competent for those zones. Germans with allied passes were permitted to use allied trains; this practice was questioned by the Soviets as early as 1945, though tolerated by them.

This then was the Berlin regime originally agreed to by the four powers. Its limitations would become evident, but it worked adequately at the outset.

6.1.1 Collapse of the Original Regime

From 1946 on, frictions between the three Western allies and the Soviet Union over governing defeated Germany steadily mounted. The disputes focused mainly on reparations demanded by the Soviet Union from the Western zones, especially the industrial Ruhr, and on the broader question of whether all of Germany was to be governed at subsistence level or helped to recover economically. Western powers wanted recovery in the interest of promoting the general economic recovery of Western Europe, for which the coal of the Ruhr was essential. More fundamentally, the Western governments feared that unless the Western European economy could be pulled out of its postwar low, national communist parties might come to power and the Soviet Union would gain decisive influence over Western Europe.

As the Soviets continued to press for unified administration of all of Germany by consensus among the four wartime allies, the Western allies responded by insisting on prior agreement among the four on a program for administering the German economy. The program was to include currency reform to eliminate the huge wartime accumulation of paper money and to motivate reconstruction, a limit on German reparations, free movement of persons and goods throughout Germany, and approval of an overall export—import plan for Germany as a whole.

In February 1948 the Western Powers and the Benelux countries, meeting in London, agreed to establish a unified German political administration of the three Western zones of occupation and to associate them with the Marshall Plan. As their meeting proceeded, the Czechoslovak communists took over the Prague government. On March Marshal Vasili Sokolovsky, Soviet representative on the Allied Control Council, stated that the London Conference of the Western Powers had decided questions directly within the competence of the Control Council. Consequently, these actions had destroyed the interallied agreement on the control mechanism in Germany. Sokolovsky walked out of the Control Council meeting. On March 30, the Soviets informed the Western powers of new controls on Allied road and rail traffic to Berlin: Allied soldiers and Allied civilian employees and their dependents travelling to Berlin by road and rail would have to identify themselves, and their belongings would be subject to check; freight shipments needed a Soviet permit. The Allies responded that documents of travelers by car would be checked and they would provide passenger lists for trains, but not individual documentation; personal property should not be inspected. In April, the Soviets boarded Allied trains and attempted to inspect documents of individual passengers proceeding to Berlin. The Allies stopped their own passenger train service to Berlin rather than subject themselves to these controls. Also in April, the Soviets closed two of the three lines used by German freight between the Western zones and Berlin, and barge traffic was subject to long delays. Despite these increasing warnings of a tightening noose around Berlin, the United States was not moved to make an explicit commitment to remain in West Berlin. Its failure to do so may have weakened deterrence and facilitated Stalin's decision to clamp down in June.

On June 18, 1948, the Western allies carried out a currency reform for the Western zones of occupation. On June 19, the Soviets completely stopped rail and road traffic from the West to the Western sectors of Berlin. On June 24 and 25 all incoming and outgoing truck traffic and postal services to Berlin were stopped, as well as food deliveries from the Soviet zone. On July 1 the Soviets left the Allied Kommandatura, which governed Berlin, as they had already left the Allied Control Council itself. In doing so, they stated that there was "no longer an Allied Kommandatura." The Western airlift began on June 26, 1948, and continued until May 12, 1949 with a total of nearly 250,000 flights. In September 1948, organized groups of communist demonstrators pushed the elected mayor and city council of Berlin out of the City Hall in the Eastern Sector, and in November a meeting of communist party members appointed a separate administration for East Berlin. A trickle of carefully checked individuals could cross the Berlin sector boundaries and the borders of the Soviet zone of occupation, but both city and country were divided.[4]

In its Note of July 14, 1948, replying to the protests of the Western allies over the imposition of the Berlin blockade, the Soviet Union argued that in violating wartime and postwar agreements providing for unified allied administration of Germany as a whole by moving to set up a separate government for West Germany, the Western powers had "undermined as well the very legal basis which assured their right to participation in the administration of Berlin. Berlin lies in the center of the Soviet zone and is a part of that zone."[5]

The original Soviet objective in imposing the blockade was probably mainly to dissuade the Allies from their actions in the Western zones and, if that failed, to push the Allies out of Berlin and incorporate the troublesome Western sectors in the Soviet zone of occupation. But although the Soviet Union declared that there was no longer any legal basis for the continuation of Allied forces in the Western sectors of Berlin, it did not in fact urge the withdrawal of these forces or intimate that it would try to expel them. Whatever the original motive, we may assume that the circumstances surrounding the Berlin airlift—its role in mobilizing political and public opinion in the Western sectors of Berlin, in West Germany, and in the Western countries—finally convinced the Soviets that it was not feasible to expel the Allies from Berlin. Consolidation of the Soviet hold over the Eastern Sector of Berlin and the Soviet zone of occupation did take place.

The Soviet Union first made known its willingness to lift the Berlin blockade in Stalin's January 30, 1949, interview with Kingsbury Smith in which Stalin indicated willingness to lift the blockade without requiring, as in earlier statements, that the Allies cancel their currency reform for Berlin.[6]

---

[4] The documents cited can be found in Wolfgang Heidelmeyer and Guenter Hindrichs, eds., Documents on Berlin (Munich: R. Oldenbourg Verlag, 1963), pp. 5, 14, 23, 31, 36, 39.

[5] Ibid. pp. 53–78.

[6] Ibid. pp. 72, 73.

6.2 The Post-Blockade Berlin Regime

The meetings of the foreign ministers and ensuing discussions among representatives of the Four Powers in Berlin restored access to Berlin. But they did not reestablish the Four Power Allied Control Council or the Allied Kommandatura for Berlin. Instead, they established a new Berlin regime for a divided city in a divided country, one in which both the Soviet Union and the Western allies relinquished, for a time at least, their own hopes for a conclusive solution of the Berlin problem, either through incorporation of the Western sectors in the Soviet zone of occupation, or in reestablishing a unified Germany where Berlin would once again become the capital city.

The new Berlin regime, superimposed on radically divergent views about its legal underpinnings that persist to this day, represented the continuation of some pre-blockade procedures and the modification of others. The fact that the Soviet Union was increasingly frustrating the exercise of joint Four Power rights in the Eastern Sector of Berlin did not eliminate these rights. Indeed, the Western powers insisted on retaining remnants of the earlier Four Power system that had survived the blockade: the Berlin Air Safety Center and the circulation of Western military patrols in the Soviet sector.[7]

Yet, on access to Berlin, there were significant differences from the pre-blockade regime. At Soviet order, instead of the earlier West German personnel and locomotives, the East German railroad system operated the Allied military trains to and from Berlin. No Germans were permitted on Allied trains, either as employees or as passengers travelling on Allied passes. The number of permitted trains was reduced. Lists of crews of West German barges had to be submitted for approval to East German officials operating under Soviet control. There was in fact tighter Soviet control over access to Berlin than before the blockade.[8]

Even more significant for the further development of the Berlin regime was the establishment of separate governments in West and East Germany. In May 1949 a constitution for each of these German governments was approved. Significantly, both constitutions laid claim to Berlin, declaring it respectively to be a *Land,* or state, of the Federal Republic of Germany and the "capital city of the GDR."

The establishment of separate governments not only formalized the division of Germany and of Berlin, it began a process of progressive 'Germanization' of the Berlin regime on both sides. On the Eastern side, this process took the form of progressive devolution to East German authorities of the Soviet Union's responsibility for its occupation sector in East Berlin and for control of access to Berlin other than access of the military forces of the Western allies. Henceforth, in the Soviet and East German vocabulary, there was only one Berlin: "Berlin, the capital of the GDR," as stated in the GDR constitution. 'Berlin-West' was an area

---

[7] Foreign Relations of the United Slates 1949, vol. III (Washington, D.C.: U.S. Government Printing Office, 1950), pp. 666–667.

[8] Ibid. p. 77.

on the territory of the GDR temporarily occupied by the three Western powers. In 1953 residents of East Berlin were issued GDR personal identity cards. In 1954 responsibility for foreign trade and foreign affairs involving East Berlin was transferred to GDR authorities.

In September 1955 the Soviet Union and the GDR signed a treaty ending the residual occupation regime for the GDR. The treaty followed similar action by the Western allies in the Paris Agreements of May 1955 to end the occupation regime in the Federal Republic. The Soviet-GDR treaty specified that both the Soviet Union and the GDR retained those obligations under international agreements that concerned Germany as a whole. The treaty was accompanied by an exchange of letters between GDR Foreign Minister Lothar Bolz and Soviet Deputy Foreign Minister Valerian Zorin in which the Soviet Union formally conferred on the GDR responsibility for supervising the border of the GDR with the Federal Republic, the surroundings of Berlin, the sector crossing points in Berlin, and communication routes between West Berlin and the Federal Republic. However, the letter explicitly reserved to the commander of the Group of Soviet Forces in Germany operation of controls over movement of Allied troops between West Berlin and the Federal Republic "for the time being, until a pertinent agreement is reached."[9] The GDR was given responsibility for regulating and controlling German civilian traffic to and from the Western sectors of Berlin. Henceforth, in the event of frequent hindrances by the GDR of civilian traffic to Berlin, the Western allies would protest to the Soviet Union, which disclaimed all responsibility.

The Western allies protested this Soviet action too, stating that the Soviet Union could not unilaterally free itself of earlier Four Power obligations concerning communications between various parts of Germany, including Berlin.[10] But as we have seen, among these earlier Four Power obligations, there was no written commitment by the Soviet Union to assure the access of civilian goods and persons from West Germany to Berlin and the protests remained without effect.

There were major interferences with German civilian access to Berlin in the fall of 1949 and several times in 1951. The Federal German government made clear to the GDR in concluding the 1951 agreement on interzonal trade between the two Germanies that the favorable terms in that agreement could not be repeated if there was interference with movement of persons and goods to Berlin. The outright interferences stopped in the fall of 1951 after a threat by the Federal Republic to stop all interzonal trade, but in September 1951 the GDR nonetheless enacted a stiff road use toll for traffic to Berlin, which was doubled in 1955 and had to be negotiated downward.[11]

In their weaker beleaguered situation in Berlin, the Western allies were deliberately punctilious with regard to their actions associating the Federal Republic with the operation of the Western sectors. In order to avoid undermining

9 See Chap. 2 of manuscript of forthcoming book on Berlin by James Sutterlin and David Klein.

10 Documents on Berlin, p. 174.

11 Ibid.

their own rights, already directly contested by the Soviet Union, the Western allies suspended application to Berlin of Article 23 of the Basic Law of the Federal Republic, which declared Berlin to be a *Land* of the Federal Republic. They declared that Berlin was not a part of the Federal Republic and not governed by it. Yet the Western allies supported the establishment of close economic ties between the Federal Republic and the Western sectors. In December 1949 the Federal Republic obligated itself at Allied request to give the Western sectors the largest possible economic support. By January 1952, through Bundestag approval of the Third *Ueberleitungsgesetz,* the Western sectors were fully integrated in the economic, legal, finance, and customs system of the Federal Republic. Nearly every law and ordinance adopted by the Federal Republic was applied in Berlin, but only after review by the three Western Allies, who reserved to themselves the final decision in all matters of status and security.

By the end of 1952 Bonn had also committed itself to include the Western sectors in all international agreements it concluded, except for those involving allied reserved rights. Here too, the Allies review all international agreements before application in Berlin, and occasionally suspend their provisions. In order to avoid justified Soviet complaint over deviations from the agreed status of Berlin, the Allies maintain in effect in the Western sectors postwar Four-Power ordinances regarding the demilitarization of defeated Germany that prohibit all organized military activities, including those of the Federal German forces in the Western sectors and conscription for the Federal German Bundeswehr, and the production there of military equipment of any kind. They regularly, if vainly, protest public activities of the GDR armed forces in the Eastern Sector on the grounds that the Eastern Sector too must be kept in fully demilitarized status in compliance with the original Four Power decisions.

To bolster the morale and the economic recovery of Berlin, many agencies of the Federal Republic were located in the Western sectors: technical institutes for physics and for testing of physical substances, the government agency responsible for supervising bank credits, and the Administrative Court. Twenty-two Berlin deputies belong to the German Bundestag. These deputies are prevented by Allied restrictions from voting to enact legislation. But Berliners can and do preside over the *Bundesrat,* the upper chamber of the Federal German parliament, belong to the Federal German cabinet, and vote for the election of the Federal President in the German *Bundesversammlung* or Federal Assembly, composed of the members of the *Bundestag* and of the *Land* legislatures of the Federal Republic.

In order to symbolize the closeness of Bonn and Berlin, the Federal Republic deliberately developed an important "demonstrative presence" of the Federal German parliament and government officials in the Western sectors: *Bundestag* committees, the *Bundestag* in plenary session, the *Bundesrat,* and the *Bundesversammlung.* "Work weeks" of the committees of the *Bundestag* took place in Berlin, as did meetings of the *Bundestag* caucuses and national conventions of the political parties represented in Berlin.

Interestingly, the GDR welcomed these demonstrative ties in their early stages. The President of the GDR parliament, the *Volkskammer,* sent the members of the first Federal Assembly in 1954 an official letter of greeting. *Neues Deutschland,* the official newspaper of the East German Communist party, welcomed the first plenary session of the Bundestag in 1955.[12]

But this indulgent Eastern attitude disappeared in 1958 with a new crisis over Berlin.

6.2.1 The Crisis of 1958–1962: The Berlin Regime Under Pressure

To deal with European and American concerns following the portentous 1957 success of the Soviet Union in launching its Sputnik satellite and developing the first ICBM, the Eisenhower government proposed deployment of American medium-range Thor and Jupiter ballistic missiles in Europe. In December 1957 the NATO Council decided to accept this proposal. In addition to American bombers already stationed in Europe, the new missiles could strike Soviet territory more rapidly (and in far greater number) than Soviet missiles could reach the United States. The American missiles were stationed in Great Britain, Italy, Turkey, and Greece. Like American Pershings and cruise missiles 20 years later, they were also intended to be deployed in the Federal Republic.

Then, as later, the Soviet Union reacted with great concern to the prospect of deployment of intermediate-range American missiles in Germany. As Jack Schick puts it in his book on the 1958 Berlin crisis, that crisis began with a missile crisis—the project for American IRBM deployment in Europe—and ended with a second, even more important one—the Cuban missile crisis. In the settlement of the Cuban missile crisis, the United States withdrew its IRBMs from Europe.[13] In the event, the American IRBMs were never stationed in Germany; Chancellor Konrad Adenauer took Soviet protests seriously and, in the interim, the United States had successfully tested its own ICBMs.

Prior to the December 1957 decision of the NATO council to deploy the missiles, the Soviet Union had attempted to forestall their deployment in the Federal Republic by proposing a confederation of the two German states and adoption of the Rapacki plan for a zone in Central Europe free of nuclear weapons.

The new Berlin crisis began on November 27, 1958, with Soviet notes to the three Western powers and the Federal Republic. The Soviet government now declared the London Protocol of September 12, 1944, null and void and demanded the termination of the 'obsolete' occupation of Berlin, either by transferring the Western sectors to the sovereignty of the GDR in whose territory the Western

---

[12] Gerhard Wettig, Das Vier-Mächte-Abkommen in der Bewährungsprobe (Berlin: Berlin Verlag, 1981), pp. 48–49.

[13] A good short account of the development of Bonn-Berlin ties is contained in Die Berlin-Regelung (Bonn: Press-und-Informationsamt der Bundesregierung, 1971), pp. 218–234.

sectors were located, or, as a compromise, by converting West Berlin into a demilitarized free city. If this latter compromise solution were not carried out during a 6-month grace period, the Soviet Union would withdraw from its remaining occupation functions for Berlin and for access to it and transfer them to the GDR, with which an agreement providing for transit to the free demilitarized city would have to be negotiated by the Western allies.[14]

As in the 1948–49 Berlin crisis, the Soviets again seemed to be testing for two objectives. The first was to use a crisis centered on Berlin to bring the Western allies and the Federal Republic to refrain from actions that the Soviets opposed—in this case, deployment of American IRBMs in Western Europe and especially the Federal Republic. The second objective was to push the Western allies out of the Western sectors. By 1961 a third objective was to limit the flow of East German refugees through Berlin, the most porous point in a tightening system of East German border controls intended to plug the economic drain caused by the loss of many educated professionals and trained workers. Whether the Soviets had serious expectations that the Western allies could actually be pushed out of Berlin is unclear. But, as his later actions in bringing Soviet missiles to Cuba showed, Khrushchev believed that the Soviet system was increasingly powerful and successful, that the Western system was in decline, and that pressure could pay off. In any event, the Soviet intention to challenge and to change the existing Berlin regime was clear.

In his tough meeting with newly elected President John F. Kennedy in Vienna in the summer of 1961, Khrushchev threatened a separate peace treaty with the GDR that would eliminate the occupation regime in West Berlin. If the United States insisted on exercising occupation rights after the new Soviet-East German peace treaty, and if East German borders were violated, Khrushchev said, "force would be met by force—if the United States wanted war, that is your problem."[15]

On his return to Washington, Kennedy responded with a partial mobilization of American forces, reinforcement of U.S. forces in Germany, and the development of plans for a military probe on the autobahn to Berlin. Widely publicized Soviet and GDR threats to cut off access to Berlin resulted in a sharp increase of refugees from East Germany, now totaling up to 2,000 a day. On August 13, 1961, the Soviets and GDR did cut off East Berlin from the Western sectors with barbed wire fences and they began the construction of the Berlin Wall. Elevated railroad and subway connections between the Western sectors and East Berlin were severed, residents of East Berlin were forbidden to visit the Western sectors, and West Berliners to visit East Berlin or East Germany. Postal and telephone connections had already been cut in June and July.

On August 18, the United States sent a combat unit up the autobahn to Berlin. But the Western powers did not carry out recommendations by some officers to challenge the Wall by military measures. They protested against the Wall, but

[14] Jack M. Schick, The Berlin Crisis 1958–1962, p. 233.

[15] Text in Documents on Berlin, pp. 180–196.

accepted its existence, to the tremendous disillusionment of Berliners and Germans generally. A turning point in German thinking about the division of Germany had been reached, with effects that were not long in coming. But the Soviet and East German actions had changed the Berlin regime immediately, by sealing off the Western sectors from their geographic environment.

The next and final act of the Berlin crisis came after the Cuban missile crisis and its denouement in October 1962. In a speech in December 1962, GDR Party Secretary Walter Ulbricht said "peaceful co-existence is linked up with compromises on both sides" and suggested negotiations between the two German states. Khrushchev in a January 1963 speech to the East German Communist Party in Berlin said that the Soviet Union had already gained its main point through establishment of the Berlin Wall and by calling the Western powers to order by stepping on their 'pet corns.'[16] He intimated that, for the time being, he would not press his demands for a change of status in Berlin. The Cuban crisis seems on the one hand to have made the Soviet leadership more sensitive to the risks of continuing the confrontation over Berlin. On the other hand, President Kennedy's indication to the Soviets, as part of the resolution of the crisis, that American IRBMs would be removed from Europe removed what appears to have been a major part of its original motivation. Finally, there had been no effective Western challenge to the construction of the Berlin Wall, which would continue to serve its function of preventing mass exodus from the GDR.

6.2.2 *After the 1958–1962 Crisis*

In June 1964 the Soviet Union and the GDR concluded a Friendship Treaty. In a joint communique accompanying the treaty, West Berlin is described as an "independent political entity" in the center of the GDR and lying on its territory. The GDR in an accompanying memorandum said that it was proceeding from the present real situation of West Berlin: that West Berlin, which in law belonged to the territory of the GDR, was still occupied by foreign powers.[17]

Khrushchev had complained in a 1959 article in *Foreign Affairs* that the real cause of tensions in Berlin was that the Western powers had progressively permitted Federal Germany to establish so many links to the Western sectors that the resulting situation had undermined the position of the Western powers. Following the wind-down of Khrushchev's Berlin crisis and Khrushchev's removal from office in October 1964, the Soviet Union suspended its pressures on the Allied position in Berlin. Instead, the Soviet Union and GDR shifted their sights to this different target, the Federal German presence in Berlin and Federal German ties with the Western sectors. They attacked the 'illegal' practices of the Federal

---

[16] Arthur M. Schlesinger, Jr., A Thousand Days (Boston: Houghton Mifflin, 1965), p. 374.

[17] Documents on Berlin, pp. 342, 351.

Republic in the Western sectors and the GDR deliberately used its control over access routes to Berlin to register its objections to these demonstrative acts of presence. Through interventions in third countries, the Soviet Union actively contested coverage of the Western sectors in international agreements concluded by the Federal Republic.

The Federal German *Bundestag* planned to hold a plenary session in Berlin on April 7, 1965, as it had in previous years. The GDR prohibited *Bundestag* deputies from using the access routes to Berlin. The Warsaw Pact held large-scale maneuvers in the area between the Western sectors and the Federal Republic, cutting off access completely six times for several hours at a time. The processing of papers for truck transport to Berlin was slowed down, often lasting a day and a half, and huge backups of waiting trucks formed at the crossing points. Canal traffic to Berlin was suspended for a week. On the day of the *Bundestag* session, which took place as planned, Soviet and GDR military aircraft flew through the Berlin air corridors, causing repeated sonic booms over the city.

In January 1968 Soviet Ambassador Semon Tsarapkin presented the Federal German government with a paper summarizing Soviet objections to Federal German activities in Berlin. The memorandum listed as objectionable practices: "Parliamentary Weeks" of Bundestag committee meetings and cabinet sessions in Berlin; the establishment of Federal agencies in Berlin; the execution of official functions by Federal German gov. eminent leaders while in Berlin; the election of the Berlin mayor as President of the Bundesrat; and meetings in Berlin of Federal German political parties and national associations. This time, the Soviets did not urge that Berlin become an independent entity. Instead, they demanded that the Federal Republic respect the "existing status" of the city. Soon thereafter, a semi-official Soviet publication criticized the inclusion of the Western sectors in the Federal German legal, economic, finance, trade, and customs system and the representation abroad of the interests of the Western sectors by the Federal Republic.[18]

6.2.3 The Run Up to the Berlin Negotiations

In the middle and late 1960s, the rigid confrontational stance of the Cold War began to erode in both East and West. Khrushchev, who had begun in the summer of 1963 to move toward improved relations with the West, was deposed in October 1964 and replaced by Leonid Brezhnev and Aleksei Kosygin. After a period of initial uncertainty as the new leadership consolidated its internal position, it continued this movement and engaged in active efforts to gain formal Western acceptance of the postwar status quo in Eastern Europe, an effort documented in

[18] Documents on Germany, 1944–1970, Committee on Foreign Relations, United States Senate, May 17, 1971, pp. 654–657.

the 1966 Bucharest Declaration of the Warsaw Pact calling for a conference on European Security. In Federal Germany, the Evangelical Church published in 1965 a significant memorandum calling for reconciliation between East and West, including East and West Germany. In late 1966 a new broad coalition of Christian Democrats and Social Democrats, led by Kurt Kiesinger as chancellor and Willy Brandt as vice-chancellor and foreign minister, succeeded the Christian Democratic-Free Democratic coalition that had governed Germany under Chancellor Adenauer and then Chancellor Ludwig Erhard for the previous 15 years. In March President Johnson and Premier Kosygin met at Glassboro, New Jersey. In October, Johnson made a significant speech on East–West reconciliation and pressed for opening SALT talks with the USSR. In December 1967, the NATO foreign ministers adopted the Harmel Report calling for a two-track NATO policy of defense and dialogue with the Warsaw Pact and, in July 1968, the Non-Proliferation Treaty was signed in London and Moscow; one of its most important provisions was to extend to the Soviet Union Federal Germany's earlier commitment to the West not to have nuclear weapons. In August 1968 East German leader Walter Ulbricht offered to hold formal negotiations with the Federal Republic on postal communications and transport issues.

The clear trend toward an East–West thaw stopped abruptly in August 1968, with the invasion of Czechoslovakia by 70,000 Soviet troops. But the momentum toward detente was strong. President Johnson continued to press for the opening of the SALT negotiations. The Soviet Union had received a decisive blow to its international standing and to its authority in the international communist movement as a result of its military repression of Czechoslovak moves toward pluralism. It made energetic efforts to regain its international repute.

Although East and West appeared to be moving toward a somewhat improved relationship, the Western position in Berlin continued to deteriorate as the GDR took firmer hold over the access routes to Berlin and gained in international stature.

In March 1968 the GDR issued an ordinance forbidding Federal Ministers and senior officials to enter the GDR and to use transit routes to Berlin. And in June, the GDR introduced the requirement that all travelers from the Federal Republic to the GDR and to the Western sectors other than members of the Allied garrisons must have GDR visas. A special tax was levied on freight shipments to the Western sectors. After protesting to the Soviet Union, the Western allies and the Federal Republic accepted the implementation of the new requirements. All pretense of Four-Power responsibility for civilian access to Berlin had been eliminated.

In the same period, the GDR itself was moving rapidly toward international recognition and the vigor of Federal German opposition to this trend was markedly declining. Up to the mid-1960s, the Federal Republic and its allies had expended great effort in maintaining the Federal Republic's claim to be the sole valid representative of the German people in the international arena and in holding back steadily more numerous GDR efforts to gain membership in international organizations. But the continued existence of the GDR and successful Soviet efforts to maintain the Soviet hold over Eastern Europe—through repression of uprisings and strikes in East Berlin, Poland, Hungary and Czechoslovakia—together with the construction of the Berlin Wall and the steady growth of Soviet military power, especially in nuclear arms, had convinced most West Germans that the division of their country was a lasting one that the Western alliance was powerless or unwilling to contest. Every year of continued existence of the GDR made it more difficult to insist on its international non acceptance.

Between May 1969 and May 1970 the GDR, previously recognized only by communist countries, was recognized by ten noncommunist countries (Algeria, Cambodia, Central African Republic, Congo Brazzaville, Iraq, Somalia, South Yemen, Sudan, Syria and Egypt).[19] Under the policy known as the "Hallstein doctrine," the government of the Federal Republic had declared it would break relations with any country that recognized the GDR and had actually broken relations with Yugoslavia (1957) and Cuba (1963) when these two countries recognized the GDR. But in the face of the many acts of recognition in 1969 and 1970, the Federal Republic contented itself with recalling its ambassadors from Cambodia and a few other countries for consultation.

Indeed, at this point, political and public opinion in the Federal Republic was moving very rapidly towards acceptance of the GDR as a separate independent country on German soil instead of as "the Soviet Zone of Occupation" or the "so-called GDR," as it was still termed at the time. In April 1969 the national convention of the Social Democratic Party, meeting in Bad Godesberg to prepare for the *Bundestag* elections of September 1969, adopted a resolution stating that the existence of the GDR as a state could not be denied. The national convention of the Free Democratic Party adopted in its election platform the demand that the Federal Republic should seek admission to the United Nations even if this meant that the GDR would gain full international acceptance by also becoming a member. Only the Christian Democrats still categorically refused to have any dealings with the GDR, making nonrecognition of the GDR a theme of the CDU campaign for the 1969 *Bundestag* election.

Senior Federal German officials made quite clear their view that further crystallization of German public and political opinion in favor of acceptance of the GDR was inevitable; they urged that the FRG and Western powers should, while it was still possible, try to gain whatever negotiating advantage they could from the GDR by dropping their opposition to its international acceptance.[20]

These forecasts were borne out after the September 1969 elections and the resulting establishment of the Social Democratic-Free Democratic government of Willy Brandt and Walter Scheel. In his October 28, 1969, speech presenting the program of the new government, Chancellor Brandt spoke of "two German states in one nation." Brandt offered a binding contractual agreement with the GDR on non-use of force and said the FRG would not attempt to block GDR trade and

[19] Gerhard Wettig, Das Vier-Mächte-Abkommen in der Bewärungsprobe, pp. 78–79.

[20] Hans-Adolf Jacobsen, Gert Leptin, Ulrich Scheuner, and Eberhard Schulz, Drei Jahrzehnte Aussenpolitik der DDR (Munich: R. Oldenbourg Verlag, 1979), p. 857.

cultural activities in third countries. In the debate on the government program, Brandt and Foreign Minister Scheel abandoned the Hallstein doctrine. Challenged by CDU speakers, they repeatedly refused to state that the FRG would as heretofore consider third country recognition of the GDR as an unfriendly act and break relations with that country.

Under the impact of these developments, the position of the beleaguered Western allies on Allied rights and responsibilities for Berlin was weakening. The Soviet Union had progressively withdrawn from its responsibilities for East Berlin and the GDR had taken on all aspects of controlling Berlin access except for access of allied troops. At the same time, the GDR was moving toward wider international acceptance. The prospect was that the Western allies would, step by step, be obliged to deal on access to Berlin, quite possibly including access of their own forces, with a German Democratic Republic they did not recognize and whose relationship with the West would not necessarily contain any of the restraints that characterized the relationship of the United States and the Soviet Union as nuclear powers. Instead, the GDR, suspected of having instigated on its own many of the recent disturbances involving access to Berlin, could continue to do so or act as the surrogate of the Soviet Union, and the Allies might have no rejoinder but direct force. Behind the Western concern over the GDR was recognition of the growing strength of Soviet nuclear arms, and of the parallel improbability that Soviet control over Eastern Europe could be seriously shaken or—considering the intervention in Czechoslovakia—would evolve to any major extent in the foreseeable future.

Moreover, the morale of the Berlin population was not impervious to the continual difficulties in access of civilian personnel and goods to Berlin. Berliners were apprehensive about growing GDR control over the access routes and Berlin production and trade suffered because of delivery uncertainties. The temporary opportunity to visit relatives and friends in East Berlin and the GDR by means of Christmas passes negotiated by Willy Brandt as mayor of Berlin in the years 1963–1966 had not recurred since Brandt's departure from Berlin to become foreign minister and chancellor in Bonn. The Soviet Union had for years refused to even discuss Berlin access with the Western powers. The Berlin regime appeared seriously threatened.

In the customary meetings of British, French, American, and Federal German foreign ministers on the margins of the NATO ministerial sessions at Reykjavik in June and at Brussels in December 1968, Foreign Minister Brandt urged his Western colleagues to undertake soundings with the Soviets to improve the situation with regard to access to Berlin and easing of visit and travel restrictions for Berliners. The American Embassy in Bonn urged that the Western allies seek to gain some advantage from the efforts of the Soviet Union to reestablish itself internationally after the serious political damage it had received from its invasion of Czechoslovakia; the Soviets should be obliged to give concrete evidence of their intentions through agreement to improvements in Berlin.[21]

---

[21] Author's own notes.

Despite these suggestions, the foreign ministries of the three Western powers were reluctant to open discussion with the Soviet Union on the complex Berlin regime. They feared that the Soviets would make far-reaching proposals for change which would have to be rejected, leaving the Berlin situation worse off than before. It took over a year from Western agreement at Reykjavik to look into the issue before the first Allied working paper for possible soundings with the Soviets over Berlin was ready.

President Nixon had visited the Western sectors in February 1969 amid the controversy over the decision to hold the next Federal Assembly there on March 5. In a speech at the Siemens plant, he signaled willingness to make changes in the Berlin situation, which, he said, was not ideal. Henry Kissinger reports that, following President Nixon's return to Washington, Ambassador Anatoly Dobrynin indicated to him Soviet willingness to negotiate on the Berlin question and that President Nixon had expressed his own willingness in a letter to Premier Kosygin.[22]

In their discussion on the margins of the April 1969 NATO ministerial conference, the foreign ministers of France, the United Kingdom, the United States, and the FRG agreed that there should be official soundings of Soviet willingness to discuss the Berlin issue. Foreign Minister Gromyko stated publicly in a July 10 speech in the Supreme Soviet that the Soviet Union was prepared to discuss the future prevention of "complications over West Berlin." The Western allies presented their discussion paper to the Soviets in August 1969. After further preliminaries, the Four Power negotiations began on March 26, 1970, in the building of the Allied Control Authority in the American Sector of Berlin.

6.3 The Quadripartite Negotiations

Given the complexity of their subject matter, the Berlin negotiations were of short duration; they ended on September 3, 1971, after a period of some 16 months. The specific objectives of both sides were reflected in their negotiating positions. Their broader objectives, which account for the success and rapidity of the Quadripartite negotiations, will be discussed later.

At the outset of the talks, the negotiating objectives of the three Western allies were modest. They were to achieve what were called "practical improvements" in the situation in and around Berlin without damage to the structure of Four Power rights and responsibilities—that is, to the existing Berlin regime. This was to be done by assuring access of civilian goods and persons to Berlin; assuring opportunity for further development of economic, financial, and cultural ties between the Federal Republic and the Western sectors of Berlin; making arrangements for less restricted movement of residents of the Western sectors to the Eastern sector of

[22] Author's own notes, 1969.

Berlin and to the GDR; and through achieving Soviet acknowledgement in some form that Western rights and responsibilities in Berlin would continue without change.

The immediate Soviet objectives in the Berlin negotiations as assessed in the West reflected interests that the Soviet Union held in common with the government of the GDR: to advance the GDR toward international acceptance; to emphasize GDR sovereignty over its territory, including the access routes to Berlin; to reduce the Federal German presence in Berlin; and to increase the Soviet presence in the Western sectors. One Soviet motive, to preserve some part of its postwar responsibility for Germany as a whole, was somewhat at odds with the rest, but clearly played a role in the outcome of the talks.

Within the framework of the new policy of rapprochement, the Quadripartite Agreement on Berlin between the former Allies established the basis for the legal status of the divided city, for relations between West Berlin and the Federal Republic, and for access to West Berlin. This photograph shows the foreign ministers of the Allied Powers signing the final protocol of the agreement. With their signatures, the agreement, which had been negotiated back in September 1971, took effect. From *left* to *right* Maurice Schumann (France), Sir Alec Douglas-Home (Great Britain), Andrei Gromyko (Soviet Union), and William Rogers (United States). Copyright (c) Presse- und Informationsamt der Bundesregierung

The early rounds of the Berlin negotiations were taken up with presentations of the divergent legal viewpoints of the two sides—the Western view that the allies were in Berlin by original right of victory over Nazi Germany and that the structure of Four-Power rights for Berlin and Germany remained intact save for Soviet noncompliance, and the Soviet view that the Soviets had the original rights in all of Berlin whose exercise they had transferred to the Western powers for the Western sectors by agreements that the Western powers had failed to respect. In this initial stage of the negotiation, the Soviets focused on their desire to dismantle Federal German activities and presence in the Western sectors and did not respond to Western suggestions that the negotiations should focus on practical improvements in access to Berlin and access for Berliners to East Berlin and East Germany. By implication, the Soviets suggested relinquishment of the extension of Four-Power status to East Berlin and the political neutralization of the Western sectors, which they described as a center of Western political and propagandistic attempts to subvert the existing order in East Germany. To this end, they argued, the 'illegal' political activities of the FRG in the Western sectors should be ended and any arrangements of a practical nature should be based on Western acknowledgement that 'West Berlin' was located on GDR territory.

The pace of the Berlin negotiations picked up after the signature on August 12, 1970, in Moscow of the German-Soviet treaty accepting the postwar borders of Poland, the GDR, and the Soviet Union. The first written position papers of the two sides in Berlin were exchanged in September 1970.

By November 1970, the Soviets had indicated readiness to accept several improvements in travel by West Berliners to East Berlin and the GDR proposed by the Western allies. However, there was no resolution of the access issue; the Soviets were asking for further cutbacks in the Federal German presence in Berlin, including elimination of meetings of *Bundestag* committees and the removal of all federal agencies, and there was little progress on the issue of Federal German representation of West Berliners abroad.

The Western powers presented a draft agreement on February 5, 1971, and the Soviets a draft agreement of their own on March 26; the drafts had some similarities on specific points, but diverged widely as regards their statements on legal and status issues. For example, the Soviet draft suggested that arrangements could be made to facilitate visits by residents of the Western sectors to "Berlin, Capital of the GDR," language almost surely pressed by the GDR, but at that advanced stage of the talks probably considered by the Soviets as "bargaining fat." Nonetheless, the unyielding nature of the Soviet draft convinced pessimists in the West that no Berlin agreement was possible. But, by the end of May, the deputy negotiators had developed a partial common text with bracketed language and footnotes indicating continued divergences. On August 23 the four ambassadors orally agreed to a text, which was initialed on September 3. The inner-German agreement was completed in December 1971 and the Final Protocol of the Quadripartite Agreement went into effect on June 3, 1972, the day of entry into effect of the Federal German treaties with the USSR and Poland following a difficult ratification by the *Bundestag*.

6.3.1 The Outcome

The final form of the Quadripartite Agreement is very close to the Western draft agreement of February 5, 1971, at the time considered by some Western experts as a 'maximalist' statement of Western objectives. The preamble of the agreement states that the four governments, acting on the basis of their quadripartite rights and responsibilities, were agreed that there should be no unilateral change "in the area." The area is not further identified, reflecting continued divergence of view over the legal status of Berlin. The operative provisions of the agreement are: (1) A declaration by the Soviet Union that transit traffic by road, rail, and waterway of civilian personnel and goods between the Western sectors of Berlin and the Federal Republic "would be unimpeded, and would receive preferential treatment." (2) The three Western powers state that ties between the Western sectors and Federal Republic will be maintained and developed, taking into account that the Western sectors are not a constituent part of the Federal Republic of Germany and not governed by it; the agreement states that the Federal president, the Federal government, the *Bundesversammlung,* the *Bundesrat,* and *Bundestag* would not perform constitutional acts of governing the Western sectors. (3) The Soviet Union pledges improved communications and travel for Berliners to the "surrounding area." (4) The three Western powers declare they would represent the Western sectors abroad in all matters concerning security and status. Subject to this limitation, the Federal Republic could represent the interests of Berlin abroad, Berliners could participate in FRG delegations to international meetings, and international conferences could take place in the Western sectors if jointly invited by the Berlin Senat and the Federal German government. The Soviet Union accepts this statement. (5) The Western powers state that the Soviet Union could establish a Consulate General and trade office in the Western Sectors. (6) The signatories agree that these arrangements would go into effect after satisfactory agreements on access and visits between the Federal Republic and the GDR and between the Senat and the GDR had been concluded. (7) The four governments agree to consult in the event of difficulties in the application of the agreement. The entire agreement with its annexes fills only about eight printed pages.[23]

6.3.2 Evaluation

The Quadripartite Agreement was a success in the sense of achieving specific Western negotiating objectives. Each of the immediate Western objectives was agreed to: reaffirmation of Soviet responsibility for Four-Power rights and responsibilities; improved access to and from the Western sectors guaranteed by the Soviets; arrangements for visits and communications by Berliners to East

[23] Henry Kissinger, White House Years (Boston: Little, Brown, 1979), pp. 100, 407.

Berlin and the GDR; Soviet acceptance of ties between the Federal Republic and the Western sectors; a more assured role of the Western sectors in international life. The West paid something for these gains, committing itself vis-ä-vis the Soviet Union to maintain the stipulation that the Federal Republic did not govern the Western sectors and permitting a Soviet Consulate General in the Western sectors. The West failed to gain explicit mention of 'Berlin' in the agreement and there was no improvement of the Western position regarding the Soviet sector of Berlin. The West also moved far toward agreeing to full international acceptance of the GDR.

In addition to specific objectives already described, the West achieved broader objectives in the Quadripartite Agreement: reducing the risk of confrontation over Berlin, and avoiding a possible Soviet withdrawal from responsibility for Berlin that would leave the Western Powers confronted by the GDR, a nation they did not at the time recognize and over which they had no hold, and which could function, like Cuba, as a surrogate for the Soviets without directly involving the USSR unless it wished to be. The Western allies gained these benefits before it became too late to do so through progressive international acceptance of East Germany. In particular, after the lapse of 25 years, the Western allies made good the enormously important omission in the first Berlin regime, the absence of contractual agreement assuring access to Berlin of civilian personnel and goods. They induced the Soviets to drop their efforts to push the Western allies and Federal Germany out of the Western sectors, to reaffirm the 'occupied' status of the Western sectors, and to explicitly accept a Federal German role in Berlin at a certain level.

The agreement brought the GDR enhanced international status and the certain prospect of further progress. In the draft agreement on inter-German relations the GDR had given the FRG in 1968, the GDR had proposed simultaneous U.N. entry of the FRG and the GDR following conclusion of inter-German agreements, and this is what occurred. Through the inner-German agreement on transit to Berlin, the GDR also gained a lump sum transit fee for use of the access routes to Berlin. With this agreement, the GDR also received the assurance of a series of annual hard currency payments of about one billion marks, which the FRG makes annually in connection with Berlin access and services to Berlin.

During the negotiations, there were some signs of division of views within the GDR leadership over the Berlin Agreement, especially over the issue of Soviet responsibility for access over GDR territory, which limited GDR sovereignty on this issue; and over the potential effects on GDR control over the East German population of a stream of West German and West Berlin visitors. To explain how it came to assume the responsibility in the Quadripartite Agreement for unimpeded traffic routes to Berlin, the Soviet Union developed the rather artificial theory that, subsequent to its return to the GDR of authority over the access routes, it had been empowered by the GDR to act as its agent in access matters in the Quadripartite talks. In May 1971, GDR leader Ulbricht was suddenly obliged to resign and the Berlin talks moved more rapidly after that.

From the Quadripartite Agreement, the Soviet Union gained the desired advance of the GDR toward international recognition and consolidation, and reduction and limitation of the Federal German presence in Berlin. It overcame much of the onus of its invasion of Czechoslovakia and reestablished its standing in the West. But Soviet motives extended beyond these achievements and were broader than those of the West, which were more closely linked with the Berlin and German issues.

Since the mid-1960s, the post-Khrushchev Soviet leadership had been engaged in an effort to open a new political and economic relationship with the United States and Western Europe. Soviet motivation to do this had been increased by the Soviet clash with the Chinese on the Usuri River in March 1969. The Soviet leadership also probably had a direct interest in defusing the Berlin situation and thus avoiding a direct conflict with the United States over this issue. This was not least because the GDR had more than once apparently taken the initiative to cause difficulties on the access routes without full advance consultation with the Soviets.

But above all, Soviet leaders realized they could not achieve the objectives of their own Western policy without some improvement of the Berlin situation. Beginning in 1969, after the invasion of Czechoslovakia had jeopardized East–West relations, the NATO Council challenged the Soviet Union to give concrete evidence of its professed desire to improve these relations through improvements in the Berlin situation.

Stronger even than its desire to reestablish its post-Prague standing in the West was the Soviet desire to achieve Western acceptance of the postwar borders of Central and Eastern Europe. Federal Germany under Chancellor Brandt believed it was necessary to accept that status quo in order to improve it. It negotiated treaties with the Soviet Union and Poland, accepting the loss of East Prussia and Silesia and the establishment of the GDR and the resulting borders. But after sometimes heated discussion caused by the Federal German government's apparent willingness to proceed with ratification of the Eastern treaties, Federal Germany and the Western allies agreed that the *Bundestag* should not ratify these treaties prior to conclusion of a Berlin agreement.

Not satisfied with Federal German acceptance of the postwar borders, the Soviet Union also wanted wider endorsement of the borders through a Europe-wide conference, the CSCE, which finally convened in November 1972. But, as shown in successive NATO communiques in 1970 and 1971, the NATO countries also made their participation in a possible CSCE dependent on prior conclusion of an acceptable Berlin agreement. This linkage was positive linkage, associating a Berlin agreement with something the Soviets wanted to achieve. That the Soviets themselves perceived this linkage became evident. After completion in December 1971 of the inner-German agreement on transit traffic to Berlin and the Berlin Senat-GDR agreements on improvement of travel and visits and on the exchange of enclaves in Berlin, the Soviet Union applied reverse linkage and declined to put the Quadripartite Agreement into effect until after the *Bundestag* had ratified the Federal German treaties with the GDR and Poland; the entire package entered into effect the same day, June 3, 1972. Other linkage factors were the announcement of the SALT I agreement in May 1971, approval of a $500 million American loan to the USSR in connection with the Kama truck factory, and discussion of American sales of grain and oil drilling equipment to the Soviet Union.[24]

Powerful as it was, this linkage or leverage in favor of the Western participants had to move the very considerable weight of near total control over the environs of Berlin by the Soviet Union and GDR, which in the short-range sense had time on their side as the world community—and Federal Germany-moved toward acceptance of the GDR as a separate state.

6.3.3 Significance of the 1971 Agreement in the East–West Context

Through the 1971 Quadripartite Agreement, the Berlin issue, which had been both product and cause of the East–West confrontation, was deliberately decoupled from the continuing world-wide U.S.-Soviet political and military competition. In the course of decades of confrontation over Berlin, the potential costs of East–West war had grown so great that both sides were willing to insulate the Berlin issue with a new modus vivendi. For the West, the continued effort to block the GDR from international recognition was not worth the ultimate risk of conflict with the Soviet Union. For the Soviet Union, the risks of continued pressure on Berlin as a source of leverage on the West or as a long-term means of squeezing the Allies out of Berlin had become greater than potential benefits. And that pressure had become counterproductive in terms of the Soviet Union's own Western policy: to gain formal Western acceptance of the status quo in Europe, and seek a more advantageous political and economic relationship with the states of the Western coalition.

In concluding the 1971 agreement, both sides suspended their long-term objectives and ambitions for Berlin, accepting that they could not be implemented at the time. By accepting a commitment to the Soviets that the Western sectors were not a constituent part of the Federal Republic, the Federal Republic accepted continued suspension of the provision of the Basic Law declaring Berlin as a *Land* of the Federal Republic. The GDR and Soviet Union relinquished active pursuit of efforts to establish GDR sovereignty over Western sectors or to declare them an international city as a step on the way to eventual incorporation in the GDR. All six powers acknowledged that no final solution was possible at the time, only a modus vivendi or regime to administer an unresolved problem.

The Berlin regime was expanded to cover civilian access to Berlin and, largely for this purpose, further modified through associating the two states which had emerged on German territory with its operation. The Quadripartite Agreement, implemented by the Federal Republic-GDR agreement on transit and the Senat-

[24] The text of the agreement with useful commentary can be found in Honore M. Catudal, A Balance Sheet of the Quadripartite Agreement on Berlin, (Berlin: Berlin Verlag, 1978).

GDR agreement on visits by West Berliners to the GDR, converted the Berlin regime from a Four Power to a Six Power regime and established a set of rules for the interaction of the six. From the viewpoint of Western legal theory, it is a two-tier regime, with the Four Powers continuing their responsibilities at one level and the Federal Republic and the GDR associated at an implementing level. But in practical terms, the day-to-day significance of the German contribution is considerably greater than that of the original Four Powers—a circumstance that may point the way to further evolution of the Berlin regime in the future.

The complex machinery for administering the Berlin regime has become still more intricate as a result of the 1971 Quadripartite Agreement.[25] In the West, the ambassadors of France, the United Kingdom, and the United States in Bonn retain responsibility for issues affecting Berlin and Germany as a whole, and thus for operation of the Berlin regime. They meet periodically with the Soviet ambassador to the GDR, who also has responsibility for residual occupation responsibilities affecting Berlin and Germany as a whole. On the basis of recommendations and studies from the three Allied missions in Berlin, Allied policy toward Berlin is coordinated with the Federal Republic in the Bonn Group, composed of officials from the three Western embassies and the German Foreign Office. In the Western sectors, still formally governed by the three Allied military commandants, their political advisers coordinate action with one another and maintain contacts with the Berlin Lord Mayor and Senat and also with the working staff of the Soviet Embassy to the GDR. This contact is the main day-to-day East–West coordinating mechanism of the Berlin regime. Since the Western allies formally retain supreme authority in Berlin, the political advisers also maintain a general oversight over the contacts of the West Berlin city administration with GDR authorities.

In turn, the Berlin authorities also oversee a wide range of commercial contacts with the GDR. The Federal Republic and the GDR each maintain a high commissioner in the capitol of the other German state, (The Federal Republic remains unwilling to accord these officials the ambassadorial status desired by the GDR because of the adverse impact on the German unity issue of recognizing the GDR as a foreign state.) The Federal Republic maintains a federal commissioner for Berlin in West Berlin and a senior official of the Berlin Senat represents Berlin interests in the Federal German *Bundesrat* in Bonn and vis-à-vis the Federal German authorities; both liaison activities are much needed in a situation where the Federal German government subsidizes about 50 % of Berlin's annual budget. The system is intricate, but it works, and usually quite well.

Today's Berlin regime is very much a regime of shared functions. For example, ground military traffic to Berlin is still processed by Soviet military personnel. Air traffic, civilian and military, continues to operate under 1945 rules and is coordinated by the Berlin Air Safety Center mechanism established at that time. In accordance with the 1971 Agreement, civilian ground traffic to Berlin is handled by the GDR under ultimate Soviet responsibility and in cooperation with the

[25] Raymond Garthoff, Detente and Confrontation (Washington, D.C.: Brookings, 1985), p. 92.

Federal Republic, which pays large lump sum transit fees and subsidies for improvement of the land and water routes to Berlin and adjudicates with the GDR the rare cases of travelers stopped or detained by GDR authorities on the land routes. The West Berlin Senat and the GDR operate the mechanism for visits by West Berliners to East Berlin and the GDR. The Federal German role in the Western sectors has been formally accepted by the Soviet Union and the GDR. Under the two-tier system, the interests of Berlin abroad in status and security matters are represented by the Allies (a function that might be exercised in the Security Council or in peace treaty negotiations, but which has remained theoretical and unpracticed) and on the second tier, the day-to-day level, by the Federal Republic, again, a more important function in practical terms.

The FRG-GDR inner-German relationship, with Federal German credits and loans to the GDR, provides a supportive framework for operation of the Berlin regime and incentives for continued GDR acceptance of it. The relationship helps assure Soviet support as well, because the Soviet Union is keenly interested in the political and economic viability of the GDR. The Soviet policy of political, economic, and military detente toward Western Europe plays a further role in maintaining Soviet support for the Berlin regime. For the Soviet Union to undercut the Berlin regime and cause serious difficulties for the Western sectors would be inconsistent with this policy. On the other hand, if there is a major adverse change in Soviet or East German policy, the Quadripartite Agreement, primarily because of its coverage of civilian access, will provide a much clearer set of rules for crisis situations than was earlier available, and greater clarity to Western decision makers and publics as to the specific obligations that the Soviet Union or GDR would in such a situation have violated.

The Berlin agreement has proved quite satisfactory in implementation. Ground traffic to and from Berlin has increased enormously, up 300% from its 1970 level, and has moved smoothly; freedom of movement and communication of West Berliners with East Berlin and the GDR greatly increased to a total of over 3 million visits a year in the early 1980s. Federal German ties with Berlin have intensified. Yet there have been difficulties, and the Agreement has required continuing negotiation and adjustment to make it work.

Representation by the Federal Republic of West Berlin's interests abroad has been accepted in principle by Warsaw Pact states, but the experience in this area has not been entirely satisfactory. West Berlin is still occasionally handled separately from the Federal Republic in exhibits and conferences held in Warsaw Pact countries, and arrangements for Federal German consular representation of Berlin firms in Warsaw Pact states are not always honored. The Soviet Union still sometimes makes a pro forma protest when Federal German leaders visit the Western sectors, doing so for example when Chancellor Helmut Kohl accompanied President Francois Mitterrand of France on a visit to Berlin in October 1985. The Soviet Union has blocked representation of the Western sectors by the Federal Republic in the Federal Republic-Soviet Union agreements on cultural and scientific exchanges. There is still no agreement on these subjects after years of negotiation because of the Berlin issue.

Until 1985, the Soviets also held back the GDR from concluding a cultural exchange agreement with the Federal Republic for the same reason. Finally, the GDR made a side agreement, accepting token participation of West Berlin ensembles in cultural exchanges. The Soviet Union and the Warsaw Pact strongly resisted the Federal German effort in 1974 to establish in the Western sectors the Federal Office for the Environment, with responsibility for environmental issues in all of the Federal Republic. This demonstrative effort to increase Federal ties with the West has led in practice to a freeze in this area. The Soviet Union has also resisted Federal German efforts to establish offices of the European Community in the Western sectors. Nor have the Soviet Union and GDR been forthcoming in honoring the right of the Federal Republic and the Senat acting jointly to invite international bodies to meet in Berlin. The Soviet Union and the Warsaw Pact states have blocked some of these conferences and have permitted others; each case has had to be individually negotiated.

Perhaps the main negative development in implementation of the Berlin agreement was the increase in November 1973 of the obligatory exchange of currency for visitors to the GDR and its application to Berliners visiting East Berlin and the GDR. The new requirement worked out to DM 25 per day, per person, a sum that imposes considerable hardship, especially on family visits by Berliners to East Berlin and the GDR (and of course by Federal Germans as well). This requirement countervenes the obligation of the Soviet Union in the Quadripartite Agreement to improve communications between the Western sectors and East Berlin and the GDR. Through assiduous negotiations and loan guarantees to the GDR, the Federal Republic has brought the GDR to reduce the obligation for a few limited groups, such as youngsters under 16.

Other aspects of the Berlin regime require continual attention. In 1984, GDR authorities on short notice closed the Glienicke Bridge crossing point used for access to the Western sectors by personnel of the Allied military liaison missions accredited to the headquarters of the Soviet Group of Forces Germany at Karlshorst, GDR. After Allied authorities protested to the Soviets, the latter intervened and the bridge was reopened. In 1986, after U.S. authorities had accused the GDR of laxness on terrorism by permitting personnel of the Libyan mission in East Berlin to enter West Berlin to carry out bombing of a discotheque there, the GDR responded with tighter security checks and demands for new documentation on personnel of the Western allies entering East Berlin as well as of foreign consulates and military missions in the Western sectors whose documents were provided by the Western allies. After repeated representations from the West, the East Germans shut down flights for inhabitants of Sri Lanka who entered West Berlin as illegal emigrants for economic reasons, but in 1986 they continued to permit a large flow from other developing countries, until this flow too was closed off after further negotiation. This has been a tactic designed to bring Federal German and West Berlin authorities to establish controls on entry to the Western sectors rather than maintaining Berlin as a single entity through minimal controls. The GDR also hoped to gain economic concessions for piecemeal reduction of the flow and these hopes were apparently fulfilled.

Although a practical success, the 1971 Berlin agreement requires continual attention, negotiation, and adjustment. Like East–West arms control agreements and other agreements it has to cover conditions that change continually, and to regulate a continuing competition between participants seeking to accumulate small advantages in the interest of long-term gain for their position.

The Quadripartite Agreement on Berlin did not end East–West competition over Berlin, as these circumstances clearly show. Both sides have accepted that they cannot now press for what they consider a full solution. Yet the competition continues. The Soviet Union and the GDR still hope for a long-term resolution as they define it. Valentin Falin, then Soviet ambassador to the Federal Republic, told Federal German *Bundestag* members and officials at the time of the Quadripartite negotiation that, in the long term, the Soviet Union continued to hope that the Western sectors would become part of the GDR. And the policy of the Federal German government, backed by overwhelming public opinion in the Federal Republic, is insistent that the German unity question must be kept open for ultimate determination by German residents of the GDR. This objective, if ever realized, would provide for a far different final solution of the Berlin question than that envisaged by the Soviet Union and the GDR.

In all probability it will take many decades for either view to develop momentum. In the interim, the expanded Berlin regime agreed to in 1971 insulates the Berlin issue from the ups and downs of other East–West relations, provides a set of rules for managing more safely the continuing competition over Berlin's status, and protects the overall East–West relationship from the negative impact of this continuing struggle. In doing so, it provides a model for application in other areas of the U.S.-Soviet competition that have potential to ignite a general crisis. It is evident that, for the foreseeable future, the relationship between these two countries and the coalitions they head will be a competitive one. No solution of the problems posed by the Soviet Union for the United States is likely to emerge, be it achievement of decisive U.S. military superiority over the Soviet Union, internal collapse of the Soviet system, or its radical change in the direction of Western pluralism. Among the urgent needs created by a situation of continuing East–West competition is for deliberate, step-by-step insulation of the most sensitive areas of U.S.-Soviet confrontation from the general flow of the relationship, neutralizing the interaction of the two, as was done through the Quadripartite Agreement on Berlin.[26]

---

[26] See Viotti, "Berlin and Conflict Management," for an excellent summary of the Four Power coordinating mechanism.

Chapter 7
A Crisis Management Center in Europe: Institutional Arrangements (1989)

There are three main reasons for establishing a regional European crisis prevention center in the framework of the Conference on Security and Cooperation in Europe[1]:

- First, there will have to be common centers for coordinating verification, perhaps in the context of both the CFE and the confidence-building talk. Why not use them for a wider purpose?
- Second, as Warsaw Pact and NATO forces build down and political change in Central and Eastern Europe continues, it will be increasingly the case that the most plausible cause of conflict either between the alliances or between individual countries of Europe is miscalculation and erroneous estimates, leading to an escalating cycle of readiness measures and possible preemptive actions. In this situation, there is strong need for some institution which can immediately cope with low-level incidents and prevent them from growing in importance and consequence.

[1] An earlier and longer version of this proposal was first published in German in 1986 as "Ein Krisenmanagement zentrum in Europa—ein konkreter Vorschlag einer Vertrauens—und Sicherheitsbildenden Maßnahme für Wien und Stockholm", in: Hans Günter Brauch (Ed.): *Vertrauensbildende Maßnahmen und Europäische Abrüstungskonferenz*, Militärpolitik and Rüstungsbegrenzung, Vol. 9 (Gerlingen: Bleicher Verlag, 1986): 329–338. This text is based on an abstract that was published as: "A Crisis Management Center in Europe: Institutional Arrangements", in: Hans Günter Brauch (Ed.): *Verification and Arms Control—Implications for European Security —The Results of the Sixth International AFES-PRESS Conference—Part I: Abstracts and Discussions*, AFES-PRESS Report No. 35 (Mosbach: AFES-PRESS, 1990): 85–86. The permission to include this text was granted by AFES-PRESS on 6 March 2014.

H. G. Brauch and T. Grimwood (eds.), *Jonathan Dean*,
SpringerBriefs on Pioneers in Science and Practice 19,
DOI: 10.1007/978-3-319-06662-2_7, © The Author(s) 2014

- Third, the two alliances will probably continue. Yet there is a strong need for a European peace structure which will take over some of their functions and, in the very long term, possibly' supplant them. In rudimentary form, such an organization exists already, the recurrent meetings of the Conference on Security and Cooperation in Europe. What is required is to flesh out the CSCE process with permanent institutions. A common European crisis reduction center would be a good starting point for gradual institutionalization of the CSCE process.

The case for a risk reduction center depends in large part on one's perspective of the future security problems of Europe. More than ever, after conclusion of East–West agreements on reduction of conventional forces, the main risk of conflict in Europe does not seem likely to be deliberate aggression of one alliance against another. Rather, the risk is likely to stem from miscalculation and a chain of reactions moving to a conflict no one originally wanted.

The main function of a risk reduction center would not be to seek to manage a full-blown East/West crisis. Rattier, it would be to prevent minor incident and misunderstandings involving the and forces of the two alliances from developing into major crises.

One can argue for either having two separate centers, one in the CFE complex and one in GDE or for combining the two centers into a single one. On its merits alone, it is difficult to resolve this issue in favor of dual centers or a single one. However, one consideration dees seem paramount and decisive here that of the most effective contribution to institutionalizing the CSCE process.

From the viewpoint of the long-term prospect of military stability in Europe, it is logical to build up the CSCE process as the emerging, if rudimentary, organizational framework of a European security system which can form a bridge between the two alliances and a framework for the continuing process of political and economic change.

A logical place to begin with the common East-West institutions which could form part of the skeleton of such a system is in a single common verification center which can be expanded into a risk reduction center and then possibly to a fact-finding and mediation center.

At the outset, the work of such a center would be routine: notification of force activities; observer and inspection requests; discussion of compliance issues and defusing of low-level incidents, like inadvertent border crossings, before they escalate in significance. Later, other functions might be added, like fact-finding and mediation in disputes among CSCE participants, say, the treatment of the Hungarian minority in Romania.

In an extreme case of anarchy in an Eastern European country, the CSCE structure at a more senior level might be used to arrange for the forces of one or more European neutral states to temporarily and symbolically take over the law-and-order function until new elections are held. This would make it possible for the Soviet leadership to avoid the high political, costs of military intervention, while at the same time avoiding NATO military involvement.

Time is short. I hope that some of the European NATO governments will present specific proposals for the establishment of parallel risk reduction centers in both CFE and the CDE confidence-building talks—we can sort out later when and how they should be amalgamated.

Chapter 8
Conventional Arms Reduction in Europe: Past, Present and Future (1990)

8.1 The Background

Two international developments important for arms control took place in the 1960s.[1] The first development, the move to negotiate on nuclear arms control, was triggered by the shock of the Cuban Missile Crisis. This terrifying event convinced leaders both in the United States and in the Soviet Union after many unproductive discussions over arms control, that they had to negotiate seriously with one another to reduce the risk of nuclear war. The initial results were the Hot Line agreement to establish a direct communication link between the US and the USSR on nuclear weapons matters and the Limited Test Ban Treaty, both signed in the summer of 1963. These developments were followed by the Outer Space Treaty, the Non-Proliferation Treaty and the move to the SALT I negotiations, which—delayed by the Soviet invasion of Czechoslovakia—began in November 1969. By the late 1960s, the process of nuclear arms control had been launched.

The second development took place in Europe. It was the move toward East–West negotiation on reducing conventional arms forces, which also got under way in the late 1960s. This second development, too, had its beginning in a shock, the shock created by the construction of the Berlin Wall in August 1961. This event convinced German and Western political opinion that earlier hopes were unfounded that German unity and an end of the division of Europe would come through rapid collapse of the communist system. German and other Western European political leaders concluded that, if the hardships of division for the population of East Germany and Eastern Europe were to be reduced, this would have to come from a step-by-step process negotiated with the Warsaw Pact states. For that purpose, the Western countries would have to accept the East German government and the governments of the Warsaw Pact, with which they up to then

[1] This text was first published as: "Conventional Arms Reduction in Europe: Past, Present and Future", in: Institute of Foreign Affairs and National Security, Seoul Korea, in a volume: *Arms Control on the Korean Peninsula—What lessons can we learn from European experiences*? (Seoul: Institute of Foreign Affairs and National Security, 1990): 23–37. This institute does not exist any longer and this text was not protected by copyright.

H. G. Brauch and T. Grimwood (eds.), *Jonathan Dean*,
SpringerBriefs on Pioneers in Science and Practice 19,
DOI: 10.1007/978-3-319-06662-2_8, © The Author(s) 2014

had minimal relations, as equal negotiating partners. With this in mind, Willi Brandt as Federal German Foreign Minister and as Chancellor played a major role in establishing a new treaty relationship between West and East Germany, Poland, Czechoslovakia and the Soviet Union.

8.2 Negotiations on Conventional Arms

Chancellor Brandt also argued that, if the United States was going to seek to decrease the risks of nuclear war through negotiation with the Soviet Union, then the NATO countries should in a parallel move also tackle the reduction of the huge East–West military confrontation in Europe. This thought was the genesis of the NATO-Warsaw Pact negotiations on reduction of conventional forces, the negotiations for Mutual and Balanced Force Reductions.

These talks lasted from the beginning of 1973 to the beginning of 1989, 16 years in all, without any specific outcome. Yet even in the absence of a negotiated outcome, the MBFR talks proved worthwhile for both sides in many ways: First, the governments of the two alliances became accustomed to non-polemical discussion of security issues and established a practice of objective discussion of negotiating issues. This experience was valuable for the successor negotiations on Conventional Forces in Europe or CFE talks. Second, the NATO countries developed and refined the main negotiating concept used in the INF negotiations, in START, and in successor CFE talks, the idea that negotiated reductions of armed forces should be to a new equal level between the two sides, a level set somewhere below current holdings of the numerically weaker side at the outset of the talks. Third, the Western countries insisted that negotiated reductions alone could not produce the desired increase in stability between the two alliances. They argued that, to meet the objective of decreased risk of war between the two sides, reductions had to be accompanied by confidence-building measures which permitted wider observation of normal activities of the armed forces and which gave clearer warning when there was departure from these normal activities and also by constraint measures which placed agreed restrictions on the force deployments and activities of each side.

In MBFR, these "associated measures" included: pre-announcement of the entry of US and Soviet forces into the Central European reduction area; preannouncement of major exercises; invitation of observers from the other side at these exercises; and limits on the size, location, number and duration of major out of garrison activities by ground forces. They also included data exchange and stringent verification with ground and air inspection.

In modified form, these measures, already discussed and agreed in NATO, formed the basis of the Western proposals at the Stockholm talks on confidence-and security building measures, whose participants were the CSCE signatory states (members of NATO and Warsaw Pact and the twelve neutral and non-aligned states of Europe). These western proposals were largely accepted by the Warsaw Pact to form the content of the successful Stockholm Document of 1986. Soviet agreement to these confidence-building measures, with their break through provisions for on-site inspection on Soviet territory, was an important step toward convincing Western leaders that Mikhail Gorbachev, designated as general secretary of the Soviet Communist Party the year before, was serious in desiring conventional arms control.

Even where no agreed outcome was reached, continual repetition of these Western confidence-building and reduction proposals in MBFR and other forums had made the Warsaw Pact governments familiar with their content and their rationale. When overall Soviet policy changed, it proved relatively easy to agree on the measures themselves. This educational process also included years of informal discussion between Western and Soviet non governmental experts which preceded East–West negotiation on arms control. When the political atmosphere in the Soviet Union changed, the new Soviet government asked these Soviet advisers for their arms control recommendations. Many of the ideas they advanced, like those of confidence-building measures and non-offensive defense, were those which Western experts had discussed with them over the years. The years of official and unofficial discussion had a productive harvest.

These benefits were worthwhile but, as noted, the MBFR talks failed to achieve an agreed outcome. The main reason was lack of interest by the political leadership on both sides, which then saw no real prospect of easing the Cold War rivalry in Europe. The Federal Germans, who had provided the main motive power on the Western side, were then stalemated internally by disagreement between Chancellor Schmidt, who wanted an outcome, and Foreign Minister Genscher, who at that time opposed any contractual relationship with the USSR on armed forces.

The next development in arms control was wholly unexpected. It was the emergence of General Secretary Gorbachev as Soviet leader in March 1985. Here was a man who took seriously the major criticisms of the Soviet system which had been made by Soviet and foreign observers over the decades. Within a year, Gorbachev backed his new policy line with impressive moves in the INF on reducing intermediate range surface-to-surface missiles and in START. In April 1986, he proposed new East–West negotiations on reducing conventional forces in Europe. This proposal for new talks was highly opportune : In the years since the MBFR talks began, the NATO-Warsaw Pact confrontation in Europe had become still more formidable: With a combined total of 7 million soldiers and airmen, 80,000 tanks, 20,000 combat aircraft, 60,000 artillery pieces and 10,000 nuclear warheads, it was the largest and deadliest peacetime concentration of armed forces in human history, costing the NATO countries about $300 billion per year and the members of the Pact about $200 billion a year. The combined total for both sides is about 60 % of all world expenditures for armed forces.

8.3 Rapid Initial Progress in CFE

In responding to Mikhail Gorbachev's proposal of April 1986, for a new round of force reduction talks to replace the still convening but unproductive MBFR negotiations, Western leaders decided to make a very stiff test of the Soviet leadership's willingness to make real progress in conventional arms control. They proposed total elimination of the Pact's huge superiority in tanks, armored troop carriers and artillery through reduction to a new equal level slightly below that of NATO's current holdings of these armaments.

To the surprise of most in the West, the Soviet Union moved rapidly to accept these terms. In December 1988, President Gorbachev announced a sizeable unilateral reduction of Soviet forces in Eastern Europe, including six tank divisions. At the opening session of the new negotiations on Conventional Forces in Europe (CFE) talks in Vienna in March 1989, Soviet Foreign Minister Eduard Shevardnadze accepted the Western concept of reduction to a new equal level below that of NATO's current holdings of the major armaments proposed for reductions.

There was further rapid progress. In May 1989, President George Bush proposed adding to reduction of tanks, troop carriers and artillery the reduction of combat helicopters and aircraft and also reduction of United States and Soviet ground and air force personnel deployed in the Central European subregion of the Atlantic-to-Urals area to a level of men each. The Soviets had proposed reduction of aircraft, helicopters and of active duty military personnel, but NATO had at first opposed this. In January 1990, the President proposed still larger reductions of United States and Soviet military personnel in Central Europe to a level of 195,000 for each country, with a ceiling of 30,000 on United States personnel stationed in Europe. Within a month, at the Open Skies meeting at Ottawa, Minister Shevardnadze agreed to this proposal for deeper cuts in military personnel.

Conclusion of the CFE Treaty means, according to NATO figures, somewhat adjusted after discussion of definitions, reduction of over 17, Warsaw Treaty tanks as compared to 2,000 NATO tanks; over 12, Warsaw Treaty artillery pieces to 0 NATO pieces; about 17,000 Pact armored vehicles to 0 NATO armored vehicles; 1,500 Pact helicopters compared to 235 NATO helicopters. Helicopter reduction will be by conversion to a non-combat role); and, according to revised NATO proposals, about 3,000 Pact combat aircraft in return for reducing 0 NATO aircraft. Total Pact reductions are on a scale of over twenty-to-one in relation to NATO reductions; in practice NATO will reduce only tanks from existing levels and no other armaments.

A useful way to visualize this huge reduction program is that the Pact countries are called on to reduce the tanks and other equipment of over one sixty heavy armored divisions, or about 40 % of the major conventional armaments of all Warsaw Pact forces west of the Urals. Not only the amount, but the nature of these armament reductions is significant. Tanks, armored troop carriers, artillery and attack helicopters and aircraft are penetration or outreach weapons essential for attack. There is no substitute for them for the offense, although there are substitutes for them in the defense.

Other features of a CFE agreement are important. There is a ceiling of men on the size of out-of-garrison activities by active-duty or reserve forces; a ceiling is placed on bridging attached to armored vehicles, and the reduction area is divided into subzones in order to control density and to prevent force concentration in the forward areas. On-site verification, to include aerial verification, is being carefully negotiated. The main consequence of a CFE agreement along these lines would be the end of the possibility of overwhelming surprise attack smashing through NATO's geographically shallow defenses which has been the main fear of Western governments for the past forty years.

Therefore, conclusion of the CFE is a momentous development in East–West relations. But other developments while the CFE Treaty was being negotiated had at least equal long-range significance. During 1989, the Moscow-dominated communist governments of all non Soviet Warsaw Pact governments were replaced, and non-communist governments came to power in the countries of Central and Eastern Europe except for Romania and Bulgaria. By agreement, Soviet forces were being withdrawn completely from Czechoslovakia and from Hungary. German–Soviet agreement was reached on the total withdrawal within four years of Soviet forces deployed in the German Democratic Republic. Poland is negotiating with the USSR reduction of the current level of Soviet forces on its territory and will almost certainly ask for complete withdrawal of Soviet forces from Poland as soon as the Soviet withdrawal from East Germany is at an end. Of the roughly 575,000 active duty Soviet military personnel deployed in Eastern Europe at the beginning of 1989, it seems probable that, by the mid-nineties, none will remain.

8.4 End Phase of the CFE Talks

In the summer of 1990, in contrast with the rapidity of early developments, the CFE negotiations had slowed down, although only temporarily. The immediate cause was that the steady stream of Soviet concessions which had provided the motive power for earlier rapid progress in the CFE talks, as well as other East–West arms control negotiations, had dried up. The basic reason for the slowdown was the apprehensive reaction of Soviet leaders to the collapse of the Soviet position in Eastern Europe, which in practice nearly reversed the balance of power in Europe, leaving the Soviet Union on its own facing a NATO which would probably be augmented in potential strength by membership of a unified larger Germany.

As regards aircraft reductions, the main substantive area of East–West disagreement in CFE during the first half of 1990, the positions of both sides were close numerically as the CFE talks entered the end phase. Both sides were arguing for a limit of between six and seven thousand aircraft per alliance. But there was disagreement over definitions. The Soviets wanted to exclude from the reduction count about 1,500 armed and combat capable two-seater training aircraft and their entire fleet of land-based bombers assigned to the Navy, which they said they would not include in reductions or numerical ceilings unless NATO countries included their own carrier-based naval aircraft. NATO in turn refused to include naval forces in the CFE negotiations or in any form of arms control.

United States and Soviet opinion appeared to be moving in the direction of postponing reduction of aircraft to a second phase of CFE talks. Only a few European countries were arguing for a further strenuous effort to include aircraft reductions in a first CFE agreement. They were right to do so. CFE participants have agreed that combat aircraft, with their speed, flexibility and range, are a key weapon both for short-preparation attack and sustained offensives. Aircraft capable of ground attack will become even more important once ground forces are reduced; these cuts will lead to increased emphasis on mobility.

Postponing aircraft reductions to later negotiation might have meant that they would not be reduced by agreement. A compromise solution providing for aircraft reduction in the first phase of CFE would be far preferable: On the main issue, treatment of land-based Soviet bombers assigned to the Navy, the Soviet Union ultimately agreed a no-increase cap on these aircraft in return for NATO agreement to place its smaller number of land-based Naval aircraft under the same limit.

The May 1990 Washington summit meeting of President Bush and President Gorbachev resulted in an important agreement on timing of CFE. President Bush agreed to participate in the summit meeting of CSCE states that Gorbachev had proposed, scheduled for mid November 1990 in Paris, while Gorbachev agreed to the United States' view that conclusion of a first CFE agreement was a precondition for holding such a CSCE summit. This compromise provided a mutual deadline to accelerate the CFE talks and virtually assured their rapid completion once the issue of German membership in NATO was resolved.

The breakthrough on that issue came in the bilateral German-Soviet talks in the USSR in mid-July 1990, where Chancellor Kohl and President Gorbachev announced agreement on a package of mutual assurances including : (l) membership of a united Germany in NATO if it so decided; (2) prohibition against possession or production by a united Germany of chemical, biological or nuclear weapons; (3) limitation to of the active-duty military personnel of a united Germany (all services);(4) withdrawal of all Soviet forces from East Germany within the next four years; (5) prohibition against deployment of foreign NATO troops or nuclear weapons in the territory of the present GDR—this measure in practice means that the territory of the GDR will become a zone of reduced military density; and (6) German commitments to take over the economic obligations of the GDR vis-á-vis the USSR, the costs of stationing and repatriation of Soviet troops in the GDR, now set at $7.6 billion, to extend a credit of $3 billion for the Soviet Union, and to support a more extensive program of Western economic aid for the Soviet Union.

This German-Soviet agreement, supplemented by agreement in the Two Plus Four talks following day on the way in which a united Germany would confirm the present German-Polish border, means in practice that a first CFE agreement will be concluded by early November this year. It also means that as soon as a first agreement is signed, follow-on talks will be convened to agree on the details of how to limit German military personnel and the personnel of other negotiation participants.

Before turning to deal further with the possible content of follow-on CFE negotiations, we should review some of the probably shortcomings of a first CFE agreement because it would be in the Western interest to tackle them in any continuation of the CFE talks. It should be kept in mind in reading this description of shortcomings that, in comparison to the positive achievements of a CFE agreement, these gaps are secondary. Moreover, they are correctable, because the negotiating process will not stop with the conclusion of a first CFE treaty, but will continue immediately.

8.5 Tasks for Future CFE Talks

A first CFE agreement is the beginning of the builddown of the military confrontation in Europe, not its conclusion. Much remains to be done, the first CFE Treaty does not contain effective limits on military personnel. A general no-increase commitment for military personnel, of signatories is included in the treaty, but this commitment is a loose one, not based on specific agreed numerical levels. The absence of specific limits on Soviet active duty ground and air force personnel stationed on Soviet territory west of the Urals creates an important potential loophole in CFE. This omission would permit the Soviet Union on its own territory well as Western participants other than Germany to increase the manpower of ground and air force units remaining after agreed armament reductions are carried out, increasing their effectiveness in the event of conflict. For example, CFE I signatories could increase the infantry or combat engineer components of armored divisions; establish additional divisions with fewer tanks and artillery, but more non limited armaments; convert reserve divisions to active-duty divisions by adding active 'duty personnel; establish a wide range of cadre units using limited numbers of active-duty personnel, to be filled up on signal by reserve personnel; or build up the ground based infrastructure of permitted aircraft, providing a higher sortie rate in the event of conflict. The conclusion is clear: manpower reductions and limits are a necessary part of force reduction agreements in addition to reduction of armaments.

Influenced by the negative experience of the unsuccessful MBFR negotiations, which were focused on manpower reductions, the NATO states had earlier decided to omit manpower from agreed cuts because of the verification difficulties they foresaw. However, if manpower reductions and limits, focused only on personnel in organized military units, with full exchange of data on personnel, armaments, and units, and with acceptance of intrusive verification, manpower reductions and limits of manpower can be adequately verified. This loophole will probably be closed in the CFE IA follow-on talks.

The terms of reference or 'mandate' of the CFE talks agreed by all 23 participants in January 1989, called for "the elimination, as a matter of priority, of the capability for launching surprise attack and for initiating large-scale offensive action". The objective of eliminating surprise attack is being met adequately in the first CFE agreement. But a first CFE agreement will not do enough to meet the second agreed objective, eliminating capability for sustained large-scale offensives.

One major reason for this shortcoming has already been mentioned, the absence of provision for agreed ceilings on military manpower and on the number of combat units in the Atlantic-to-the Urals area. Moreover, the CFE I agreement will contain no effective measures impeding rapid forward movement of forces. It will contain no restrictions on equipment and units which enable long range forward movement of forces like mobile anti-craft units, field hospitals, pipe-laying equipment, tank transporters, or bridging equipment other than that attached to armored vehicles. And it will not restrict training of reserve personnel for mobilization, containing only one restriction with regard to the size of mobilization, on the size of exercises in the field.

Owing to pressures from arms-selling NATO member states, plus the refusal of the USA to extend measures of this kind to its own home territory, CFE will also do nothing to limit the production inside the Atlantic-to-Urals reduction area of armaments of the type reduced by agreement; these arms could be stored at production plants for later use by oversized manpower complements of organized units. In CFE IA, agreement should be sought to make newly produced arms more observable through storing them in designated locations open to air or ground observation. To achieve agreement on this by European countries and the USSR, it probably will be necessary to place arms production in USSR territory outside the Atlantic-to-Urals reduction area and in the United States under the same modest reporting requirements as inside the Atlantic-to-Urals area.

Modernization of permitted arms could be an even greater problem than at present under conditions of arms reductions where action to counter introduction of new types of weapons through the traditional means of increasing armaments will be blocked by treaty limits. More needs to be done in this field, but confidence—building measures now under discussion in Vienna would require a year's advance notification of deployment of newly developed weapons. A further confidence—building measure calls for exchange of defense budgets and their discussion. This measure would make it possible to raise questions about major modernization projects years in advance of their actual production and deployment.

Verification measures for CFE will be stringent, but they will omit permanent observers at major military installations, traffic choke points or exit-entry points into the Atlantic-to-Urals reduction area, as well as travelling go-anywhere patrols; all of these measures would be useful and should be raised in the CFE IA talks, where details of verification by aircraft which were not included in the first CFE treaty will also be discussed.

Another negative by-product of CFE I should be tackled in continuing CFE negotiations. This is the fact that, unless some further constraints are applied, military planners on both sides, no longer having sufficient heavy conventional armaments to pursue a linear defense along their entire borders, and with NATO and Soviet forces separated in Central Europe by the intervening territory of Poland and Czechoslovakia, will probably emphasize rapidly moving armored units and air-delivered forces. Both sides will be equipped with still very large holdings of weapons of this type which are, as noted, essential for the offence. Both sides are likely to employ a strategy of maneuver warfare, and a meeting engagement. Even with peaceful intent, with both sides similarly equipped with a still high number of heavy armaments indispensable for attack and with preemptive speed an important value, military stability could suffer. If either side fears the other is about to attack, the advantage will go to the side which strikes first and which gets furthest and fastest with its armored columns. There are three cures for this situation : further deep cuts in offense-oriented armaments; reduction of equipment essential to rapid movement of forces; and a defense-oriented posture in border regions, with mobile elements pulled to the rear. Agreements on measures of this kind are needed if the CFE participants are to make definitive progress toward the objective of defense dominance, which both alliances have announced as the long-term objective in the Vienna talks.

8.6 Continuing CF

The Western objective in continuations of the CFE talks should not only be to make good these shortcomings of CFE I. The overall Western objective should be to cut NATO forces to a new equal plateau with the USSR, providing force levels that NATO countries can hold with less strain and cost over the next decades, while they observe long-term developments in the USSR. At the same time, NATO's objective should be, while the opportunity continues, to use the willingness of the Gorbachev government to negotiate further builddown of the East–West military confrontation in Europe in order to create additional stability between remaining forces and to build in further barriers of delay against possible decisions of a different Soviet government to build up Soviet forces.

NATO governments should also use the CFE process to save money. With reductions of military manpower and armaments of 40% from CFE levels, European NATO countries should be able to save at least 30% of their current combined defense budgets of $150 billion a year. If, as appears probable, for the purpose of reinforcing the Gulf, United States manpower in Europe were reduced to under 150,000 from the current level of over 300,000, and the withdrawn forces disbanded, this should also bring savings of up to $30 billion a year. Consequently, the next set of CFE talks should be conceived not only as narrowly focussed on dealing with the issue of German military manpower, but as far broader in scope. But neither the Soviet Union nor the Eastern European states nor its European neutral states wish to continue the CFE talks on the current alliance-to-alliance basis. The Eastern European states do not wish to be obliged to seek common Warsaw Pact positions with one another and the Soviet Union does not want to undertake the difficult task of trying to achieve unified positions or to accept sublimits which restrict it to only 60–70% of the total of men and arms accorded to the Warsaw Pact as a whole—which therefore make it numerically inferior to NATO.

To meet these difficulties, a changed negotiating concept for continuing the CFE talks will be needed. One alternative would be to retain the CFE concept of reducing to equal levels, but to apply it in a different way: parity at lower levels would be sought between the forces of all NATO states or of a part of NATO states on the one side, and all or part of the territory and forces of the Soviet Union west of the Urals on the other side. The Eastern European states and NATO states not participating in the organized reductions would participate in the continued negotiation and would commit themselves either to make individual reductions or to accept enduring, upper limits on their force levels. This approach would combine reduction of NATO and Soviet military manpower with further reduction of the same armaments reduced in CFE I—tanks, artillery, armored personnel carriers, combat helicopters and aircraft. The objective of a CFE II of this type should be cuts of between 20 and 40% in both heavy armaments and active—duty manpower from the NATO's post CFE I levels, with Soviet forces coming down to NATO's new level, either for the wide Atlantic-to-Urals area or for a portion of it. To the extent possible, reductions should be by units, with no-increase limits on remaining units of the type reduced.

8.7 The New European Security System

The CFE agreements will not stand in isolation. They will form part of a new emerging security system in Europe. In addition to probable CFE and CFE I-A agreements, the next two years may well see an agreement on short range nuclear forces which will probably see the elimination of nuclear artillery shells and of short-range nuclear armed surface-to-surface missiles from Central Europe. There will be one or more additional agreements on confidence-and security-building measures and possibly an Open Skies agreement.

Together with the security provisions of the agreements on German unification, this large body of rules and restrictions on armed forces in Europe will form the framework of the new European security system. This European system will also be composed of a number of integrative organizations, first and foremost, the European Community, the NATO alliance, possible bilateral treaties of assurance between Germany and its.

Eastern European neighbor states, and the permanent institutions of the Conference for Cooperation and Security in Europe, which got their start in the Paris summit conference in November 1990. These institutions include a Center for the Prevention of Conflict which has considerable potentiality for useful achievements in heading off military and political crisis. At a minimum, this Center will be charged with investigating unexplained military activities. Later, it will probably be given the responsibility for operating agreed procedures for settlement of political disputes that are now under discussion among CSCE countries. The Center should, in my view, also be given responsibility for obligatory fact' finding missions in both political and military disputes.

8.8 Conclusions

What are some of the general conclusions which can be drawn from this account of negotiations on reducing conventional forces in Europe? First, the willingness to negotiate on arms control between East and West required as a prerequisite a willingness on each side to accept the other side as a valid negotiating partner. Second, formulation, presentation, justification and repetition of Western arms control proposals proved a worthwhile investment even in the face of repeated rejection by the Warsaw Pact. When the time came for the Soviet Union to move, it was these previously rejected Western proposals which in most cases provided the content of agreed measures. Third, the contribution of President Gorbachev was essential to success. His decision to change the foreign policy of the USSR and cut the Soviet defense budget and his willingness to accept intrusive verification and to accept far larger Soviet than NATO reductions in order to achieve this goal was central. The CFB talks are a classic situation of genuine arms control, where one partner has a genuine motive of its own to reduce its own armed forces and therefore desires to bring the other side's forces down along with its own. This illustrates the point that, in situations of protracted stalemate, progress can be made through confidence-building measures, but also that it is very helpful if one side takes the initiative to make far-reaching proposals. Classic arms control—reducing the armed forces of potential adversaries by mutual agreement—remains the core of what is taking place in Europe. Smaller armed forces on both sides do not of themselves bring stability. But smaller forces which have to be built up over time to invasion size are a valuable component of stability.

The agreement in CFE over what was to be reduced was also significant. Tanks, mobile artillery, armored combat vehicles, combat aircraft and combat helicopters are armaments which are absolutely essential for the offensive and for penetrating attack. All of these weapons are also useful in defense. But in defense, there are substitutes for these weapons in the form of heavy firepower, short-range rockets, anti-tank weapons, anti-aircraft weapons, minefields and defensive positions. In offense, there is no substitute for these armaments. In bringing governments on both sides to act on the basis of this insight, the European proponents of non-offensive defense made an important contribution.

Finally, the insistence that reductions be accompanied by transparency-creating confidence-building measures and constraints on military deployments and activities, plus carefully thought-out, intrusive verification, points to the need to accompany or to precede negotiated force reductions, regardless of where they are carried out, with measures of this kind.

The new security system in Europe will consist of a combination of armed forces and arms control agreements. In adversarial situations any where in the world, this combination is the rational choice for governments. It provides more security and is cheaper than reliance on armed forces alone.

Part III
Texts by Ambassador Jonathan Dean on the New Security Order in Europe since the 1990s

Jonathan Dean relaxing by playing the accordion in the garden of his home in Virginia. *Source* Private photo collection of the Dean family

Chapter 9
Components of a Post Cold War Security System for Europe (1990/1992)

Both 1989 and 1990 have been historic years in Europe for arms control and for political change, as may be the next few years as well.[1] Yet the rate of political change in Central and Eastern Europe has outpaced the speed of arms control negotiation. This gap in timing raises questions about the future role of arms control. It also raises the question of how to fit together the major components of a new post-cold war security system for Europe so that they have maximum positive effect.

Owing in large part to a continuing conciliatory Soviet position, 1989 and 1990 were boom years for East–West arms control in Europe The new NATO-Warsaw Pact (WTO) talks in Vienna on Conventional Forces in Europe (CFE) progressed rapidly after their beginning in March 1989 toward a first agreement before the end of 1990 Implementation of this treaty will eliminate the WTO's large numerical superiority in tanks, armored personnel carriers, artillery, combat aircraft, and armed helicopters and establish new equal levels roughly 10% below NATO's current holdings of these armaments. New talks on confidence and

[1] This chapter was previously published as: "Components of a Post Cold War Security System for Europe," in: "Components of a Post-Cold War Security System for Europe," in Hans Günter Brauch and Robert Kennedy, eds., *Alternative Conventional Defense Postures in the European Theater*, Vol. 2 (New York–Philadelphia–Washington–London, 1992): 45–68. It draws on a study by the author on "Components of a Post Cold War Security System for Europe". in: Horst Afheldt, Malcolm Chalmers, Jonathan Dean, Hans Günter Brauch: *German Unity and the Future European Order of Peace and Security*, AFES-PRESS Report No. 38 (Mosbach: AFES-PRESS, 1990): 127–161; and on a text co-authored with Stanley Resor contained in Kurt Gottfried and Paul Bracken (Eds.): *Reforging European Security* (Boulder: West view Press. 1990). Permission to use this text was granted by AFES-PRESS on 8 March 2014. Perseus Press has returned the copyright for Westview Press to the authors and editors. Ms. Betty Sheehan, Permissions Department, Taylor & Francis Group, New York has also granted the permission on 7 April 2014.

H. G. Brauch and T. Grimwood (eds.), *Jonathan Dean*,
SpringerBriefs on Pioneers in Science and Practice 19,
DOI: 10.1007/978-3-319-06662-2_9, © The Author(s) 2014

security-building measures to augment those contained in the Stockholm Document of 1986 also brought agreement on important new measures in 1990. The Malta and Washington summit meetings brought progress toward a bilateral treaty providing for destruction of chemical weapons and toward a START agreement providing for deep cuts in U.S. and Soviet strategic nuclear armaments, an agreement that also had good prospects for a near conclusion.

The arms control process seems likely to proceed, at least for a time. The CFE force reduction talks and the talks on confidence-building measures will be continued. Both negotiations, especially the latter, will probably achieve further agreements in the next 2 years. NATO has already committed itself to beginning talks on reducing tactical-range nuclear weapons when implementation of reduction of conventional forces begins.

In politics, the years 1989 and 1990 were equally rapid moving and even more far-reaching in their replacement of the Soviet-dominated governments of Eastern Europe. By the end of 1989, the fundamental decision of the German people, especially the population of the German Democratic Republic, to unify the two German states had taken place. Negotiation in 1990 on the internal and external aspects of unification was remarkably rapid and friction-free, although it dealt with difficult issues. Within a matter of months. Federal Germany and the USSR negotiated complete, phased withdrawal of Soviet troops from the GDR. NATO membership for a united Germany, and numerical limits on the military personnel of a united Germany. The two German states negotiated two treaties providing for the integration of the GDR into the Federal German economic, political, and legal structures. The 2 + 4 Talks between the two German states and the four wartime victors over Nazi Germany–France, the United Kingdom, the United States, and the Soviet Union—agreed in September 1990 on the postwar borders of the united Germany and on dissolution of remaining Four Power rights over Germany. German unification took place on October 3, 1990. With the first all-German elections in December 1990.

9.1 The Emerging European Security System

Some contended that these remarkable political developments had made arms control negotiation in Europe obsolete and unnecessary that the East–West military confrontation would fade away through unilateral actions of the European slates and the Soviet Union But these events had actually made arms control agreements even more necessary.

From the most elementary viewpoint, political change would be impossible if the military confrontation were still going at full blast. If the arms control process were seriously bogged down, this would suspend or slow down political change. Beyond this, negotiated reduction in CFE of both Soviet and NATO forces, with no-increase commitments on the forces of both sides, accompanied by verification and confidence-building measures, are a central component of the emerging security order in Europe.

Arms control, in the form of agreed force limitations including a limit on German personnel, to be implemented in a continuation of the CFE talks, and restrictions on future military deployments in the GDR, were a prerequisite for formal Soviet agreement to German unity. And the reductions, limits, restrictions, pre-notifications, mutual observation and verification measures provided for in these various agreements, present and future, will provide the framework of the new European security system, its rules, and regulations. These rules and restrictions will be implemented and coordinated by a number of institutions, including a continuing NATO alliance, permanent institutions of the Conference on Security and Cooperation in Europe, and also by the European Community and some subregional associations. Together, the rules and restrictions on armed forces and the integrating institutions will compose the new European security system.

Will this new security system be able to meet the challenges to which it will be exposed? To answer this question, however tentatively, we should look first at the probable challenges to the system and then more closely at its component parts.

9.2 Challenges to the European Security System

There are some residual dangers in the present situation in Europe that could develop into larger ones, and against which reasonable insurance is necessary—insurance in the form of a carefully considered combination of smaller but effective armed forces, arms control measures, and of institutions for coordinating the security interests of the European countries.

For example, the integration of a united Germany into Europe could possibly break down because of developments inside Germany or outside it. Under worst circumstances, a united Germany might successfully seek domination over the EC and lead it in unconstructive directions. Or a contest for dominance between a united Germany and an anti-German coalition with the EC could lead to paralysis and even breakdown of the community. German concentration on economic and political relations with the Soviet Union could in some circum-stances lead to cooling of the U.S.-German relationship and antagonism between the two countries.

Failure of economic reforms could lead to populist dictatorships in Eastern Europe or even the collapse of civil order in individual countries. The reemergence of nationalism and ethnic frictions in Eastern Europe could lead to local conflicts that draw in the major military powers.

Even after START and CFE agreements arc implemented, the Soviet Union will remain the strongest single military power of Eurasia and the only nuclear superpower on the continent, with a very substantial capability, if not intention, to use its armed forces for political intimidation or even aggression. As evidenced by Soviet intervention in the Baltic states and domestic developments, the Soviet

Union remains an authoritarian system where radical change of policy by decision of the rulers or by political coup is possible. Domestic violence in the USSR could increase and suddenly turn outward, a disquieting prospect in view of the Soviet Union's large nuclear arsenal. Specific short-term problems could arise as long as Soviet forces are still in East Germany.

Some believe that the Soviet Union already can be discounted as any sort of threat to the Western nations because it is in the throes of dissolution or because it is moving with certainty toward a functioning democracy. There is evidence to sustain both views, but it is far from conclusive, and the two views are partially contradictory. Soviet capabilities will have to remain the guideline for sizing NATO forces even at reduced levels.

Other dangerous contingencies may arise from unstable conditions in the Near East and northern Africa, which could seriously impede the commerce and energy supply of the European Community states and of the United States. The Gulf war has already raised this problem in a pressing way. France. Spain, Portugal, and Italy have expressed concern over the political-military effects of declining economics, increased population, and the growing strength of Islamic fundamentalism in North Africa.

Some of these developments are unlikely, but all are possible, and it is the role of insurance to cope with unpleasant developments. However, unlikely, their consequences would be extremely grave if they took place.

Above and beyond contingencies of this kind, the military confrontation in Europe will remain huge after a first CFE agreement. Each alliance will be accorded 20,000 tanks, four to five times what the Wehrmacht had at the outset of World War II The Soviet Union will be permitted enough heavy equipment in the western USSR to equip 60 active duty tank and motorized divisions, and these remaining forces will be modernized and even more effective after destruction of older equipment in the CFE regime. Well over 6 million troops will still be under arms in the Atlantic-to-Urals area even after significant Soviet unilateral reductions, deep cuts in the personnel of other Warsaw Pact states, more limited U.S. troop reductions, and major Soviet withdrawals from Central Europe.

This level of armed forces is far higher than it need be to provide European countries—or the Soviet Union—insurance against contingencies of the type we have mentioned. Beyond this, unless more is done in arms control, both the NATO states and the USSR may settle on a force posture focusing on highly mobile maneuver warfare, with force structures like that of the Wehrmacht at the beginning of World War II, creating an element of instability for the future. And the post-CFE force level will still be too high to permit NATO countries to realize real savings from their approximately $300 billion annual expenditure for defense in Europe.

Moreover, not enough has been done in the Vienna CFE talks thus far to restrict the force generation or mobilization capacity of the Soviet Union, or for that matter, of NATO. There are potential loopholes in the CFE agreement. The treaty now in prospect would not prohibit production and stockpiling in the reduction area of the very armaments that will be cut by a CFE treaty, although this production must be notified. With the exception of German military personnel in the central European sub-region, the agreement might also permit increasing the personnel of existing combat units, establishment of new units, or the complete mobilization of reserve units with active duty personnel. The CFE treaty will impose no restrictions on Soviet forces beyond the Urals, where the USSR has stockpiled large amounts of armaments removed from western Russia—beyond the destruction requirement of the CFE treaty.

9.3 Components of the European Security System

To test whether the new European security system can cope with these contingencies, we now turn to a more detailed examination of the components of that system.

The widening, step-by-step economic integration of the European states, as the Federal German economy was integrated into that of Western Europe after World War II, and the establishment of effective democratic systems in all European states, in particular the Soviet Union, are beyond doubt the best long-term guarantees for security in Europe. Their promotion, therefore, should be a priority policy aim for all Western states. The European Community, as the core of integrative dynamism in Europe, will be the chief component and foundation of the future European security system. However, in addition to a program of actions designed to promote the integration of economies and deeply anchored democracies, in order to provide adequate insurance in Europe against the contingencies described here, there will also have to be a network of mutually reinforcing security agreements and coordinating security institutions, backed by armed forces at lower levels.

In the early 1990s, these security agreements will probably include the following issues.

- A first and a second CFE treaty on reduction of conventional forces. We deal with this subject later in this chapter.
- The 2 + 4 agreement between the two German states and the four World War II allies on German unification. The security aspects of this agreement and of related agreements will also form important elements of the new European security system. They are discussed further in this chapter.
- An Open Skies agreement providing for unrestricted aerial observation over the entire territory of the USSR, United States. Canada, and of the European member states of NATO and the WTO. Although negotiations on this topic bogged down in early 1990, an agreement on this subject could be an extremely reassuring supplement to verification of arms control agreements because it would permit aircraft observers to go any-where, without being limited to monitoring specific arms control obligations.

- A treaty reducing tactical nuclear arms. As I visualize it. this U.S.-Soviet agreement, on which negotiation will begin soon after a first CFE agreement is signed, would eliminate all U.S. and Soviet nuclear-capable, ground-based missiles and nuclear artillery shells from the Atlantic-to-Urals area (AITU) and provide for a residual nuclear deterrent force of under 500 air-delivered nuclear weapons (gravity bombs and standoff missiles) on each side. A negotiated agreement is preferable to mutual unilateral reductions because it can provide for missile destruction and on-going verification of non-deployment.

NATO has proposed mutual unilateral withdrawal of nuclear artillery shells to begin when a first CFE treaty is signed. This approach has the advantage of starting withdrawals prior to negotiating a detailed agreement. However, the agreement on short-range nuclear forces should cover artillery shells to codify their prohibition in the area and to provide a basis for agreed procedures to verify the prohibition. Although nuclear artillery projectiles are relatively small and light and could rapidly be brought into the Atlantic-to-Urals area from the outside in the event of conflict, it is important that this area of verification finally be tackled between the United States and the USSR.

Although aircraft-delivered nuclear weapons too arc transportable and can be brought into Europe by both sides in the event of conflict, their peacetime presence in the area may be more reassuring to those European leaders who consider it important to maintain extended nuclear deterrence below the strategic level Deployment in Europe of U.S. aircraft-delivered nuclear armaments might bolster the willingness of these leaders to engage in cooperative measures with the USSR. Continued deployment of U.S. nuclear weapons in Europe may also make the new situation appear less of a break with the past for those Europeans and Americans worried by the far-reaching political changes in Europe. The known presence of these weapons in Europe could damp possible recklessness of hotheads on both sides. Deployment of these weapons in Europe does make it somewhat more plausible that they would in fact be used in the event of conflict and therefore would provide a more effective deterrent to conflict.

These considerations make a good case for continued but limited deployment in Europe of U.S. tactical-range nuclear weapons Nevertheless, the United States should avoid all forms of pressure on this issue. It should leave it to the government of a united Germany and other NATO governments, based on their assessment of their own security requirements in the changed situation, to reach their own decisions in this area, including deployment of tactical air-to-surface missiles on their own territory. Some governments may well reach positive decisions, especially if those decisions arc preceded by a U.S.-Soviet agreement on short-range nuclear forces, which permits each of these countries to possess and deploy a limned number of air-mounted nuclear weapons. The existence of such an agreement would mean that deployment of U.S. air-mounted nuclear arms in Western Europe would no longer be the subject of the intense East–West political controversy that such deployments have caused in the past.

In Germany, the situation has become more complicated by strong anti-nuclear feelings and German-Soviet agreement that after unification no nuclear armaments would be deployed on the territory of the former GDR. But given the fact that air-to-surface missiles are rapidly transportable and assuming that all NATO members will at a minimum permit their deployment in a conflict situation, the United States should not make the willingness of a united Germany to accept peacetime deployment of U.S. air-mounted nuclear weapons into a 'test' of the U.S.-new Germany relationship. The same goes for deployment elsewhere in Europe. It will bode ill for the future of a German-U.S. relation-ship and for the future of NATO if one of the first major international issues that confront newly united Germany is a divisive debate over nuclear deterrence or military strategy.

The issue of nuclear deterrence should recede to the background of political debate in a situation of decreased threat, especially if a U.S.-Soviet agreement on short-range nuclear forces can be achieved.

The Confidence and Security-Building Measures contained in the Stockholm Document and its continuations will be among the most important components of the new European security system. Important talks on Confidence and Security-Building Measures (CSBMs) go on in Vienna among the 35 signatory states of the 1975 Helsinki Accords of the ongoing Conference on Security and Cooperation in Europe (CSCE). The participants include all current members of NATO and the Warsaw Pact and 12 European neutral and nonaligned states, of which the most important in the military sense arc Sweden, Finland, Switzerland, Austria, and Yugoslavia. Negotiators of these states are working to augment the highly successful Stockholm Document of 1986, which contains a series of CSBMs providing for pre-notification and observation of military activities in the field. The participating governments approved a first group of additional measures in the Paris summit of November 1990 and agreed that the Vienna negotiators would continue their work with the aim of reporting further agreed measures to the CSCE review conference scheduled for March 1992 in Helsinki.

Among the most promising topics under consideration by the Vienna negotiators on confidence building measures is the question of how to further develop the discussion on military doctrine held at the beginning of 1990. Further discussions are scheduled for 1991. They could in time culminate in an agreed guideline for defense of all CSCE states, like that already familiar to NATO, to the effect that the objective of national defense should be to reestablish the status quo ante after an attack, rather than the more traditional objective of inflicting conclusive, punitive defeat on the adversary Adopted by all CSCE states and articulated in force limitations, such a code of self-restraint could eliminate the major dynamic of past competitive force buildups—the effort to gain decisive superiority in order to assure successful retaliation against a potential attacker-and would be an important component of a European security system.

A further promising measure agreed in 1990 is the exchange of data on military budgets and their annual discussion Implementation of this measure could open a prospect of slowing the rate of technological innovation in its early stages and dealing with force buildup not covered by agreements already in force. A third measure of considerable potential significance provides procedures for clarification of unexplained military incidents. This could be developed to become the core element of a useful CSCE crisis prevention regime, described further below.

9.4 Institutions for European Security: The CSCE

This framework of arms control treaties and agreements will be coordinated, implemented, and reinforced through the activities of a variety of institutions, among them, the European community, NATO, and the permanent institutions of the new pan-European security structure arising from the Helsinki CSCE process. Since adoption of the Helsinki Accords in 1975, a series of CSCE conferences has functioned as a rudimentary regional United Nations, discussing the whole range of interstate relations and reaching decisions by consensus on actions to be taken by member states. For a time at least, the agreements establishing these new CSCE institutions arc likely to have the same unanimous resolution character as the Stockholm Document of 1986, not a treaty, but considered "politically binding" by mutual agreement.

The heads of government of the states in the CSCE, in the Paris meeting in mid-November 1990, established a number of permanent CSCE institutions whose operation will be reviewed at the Helsinki meeting in March 1992 As agreed at the NATO summit of July 1990, these CSCE institutions include the CSCE Council, regular meetings of foreign ministers or their representatives, a permanent secretariat to prepare the council meetings, a mechanism to monitor elections in all CSCE countries, and a Center for the Prevention of Conflict The "Charter of Paris" of the CSCE security system contains mutual pledges to refrain from the use or threat of force in mutual relations based on the 1975 Helsinki Accords.

The states participating in the CSCE have a wide spectrum of views with regard to the scope and competence of its permanent institutions and activities, ranging from the most restrictive, represented at this stage by the United States, to the most comprehensive, represented by the WTO states as well as Federal Germany. Despite this divergence of opinions, it appeared likely that consensus would ultimately be reached that the foreign and defense ministers of CSCE states or subordinate officials periodically meeting for this purpose should have general responsibility for successful operation of CFE, confidence-building, and other arms control agreements insofar as they do not remain the responsibility of individual treaty signatories, and also for suggesting new areas of potential arms control activities. Under their supervision, a CSCE Center for the Prevention of Conflict will operate a Europe-wide data and verification center for CSBMs, coordinating data exchange, pre-notification of military activities and inspection requests arising from agreements on confidence-building measures, and probably later from CFE and Open Skies agreements. The center would also seek to illuminate and resolve at an early stage incidents involving the armed forces of participants. In cases of political disputes between member states, the center in the future might provide fact-finding, mediation, and arbitration services. In extreme cases, the CSCE governments might request some member states, such as the European neutrals, or even the United States or USSR, to provide forces to separate the armed forces of member states whose frictions threatened to break into conflict, or to establish order in the event of total breakdown of a member state government, as happened in Yugoslavia in mid-1991.

However, the purpose of a CSCE Center for the Prevention of Conflict would be crisis prevention rather than crisis management. It could not deal with an actual crisis in relations between major states, let alone a full-blown East–West confrontation. But it could investigate and clarify incidents and misunderstandings involving armed forces, in particular military alerts, which are not covered by any existing or prospective agreement, and also friction between member states over treatment of ethnic minorities. The center could inform governments of developments that carry the risk of becoming a serious crisis. A potentially extremely useful role for the center or for the CSCE Council of Ministers might be in coordinating nonproliferation efforts of all member states, including sale of missile and reactor components. The center or the group of senior officials could also gradually move toward control of arms sale to non CSCE countries; the CSCE will include all of the world's major weapons producers except China and Brazil. The process could begin without prior commitment with ad hoc discussion of continuing or incipient conflicts where outside arms supply is playing a major role.

Several related proposals may be grouped in the center: coordination of CSBM verification, a comprehensive communications system linking all member states, measures approved in an on-going project for an agreement on medication, conciliation, and arbitration of interstate disputes, and NATO's proposal, agreed at the 1990 summit, to establish a CSCE "mechanism for discussion of unusual activities of a military nature."

There is very broad support for a permanent CSCE security structure among European governments and publics. All European governments have a strong desire for common institutions that would visibly represent their determination to move toward unity after long-enforced division and their determination against further war in Europe. These sentiments are so strong that they have led to the establishment of some permanent pan-European CSCE institutions and will assure their further development. The question is how far they will develop.

Soviet and some European leaders conceive of the CSCE as a single supranational institution that would ultimately absorb NATO and the WTO and integrate their remaining armed forces in a pan-European defense community. Although the United States moved far from its original position in endorsing a permanent CSCE structure at the NATO summit of 1990, it has thus far taken a restrictive position on the issue of CSCE competence. The United States continues to consider the CSCE an actual or potential rival of NATO and takes at face value the desire of the European Left to supplant NATO and the WTO with an expanded CSCE. For this reason, the United States opposed the proposal of Federal Germany and most other European states to establish a CSCE verification coordination center, that could coordinate verification of all arms control agreements in Europe as well as verification of CSBMs Rational as such a project would be the United States feared it would undermine its effort to establish a verification center in NATO, although the two centers would in fact have different functions.

Obviously, there is a middle position between these two extremes, one of maintaining NATO and at the same time building up the CSCE and treating both institutions as complementary, as indeed they arc. It seems probable that (he CSCE consensus will form around this position.

A CSCE security system and its subordinate institutions merit strong support. But these institutions are unlikely to be as capable in maintaining European security as their proponents claim. If the larger member states disagree among themselves, the CSCE may be powerless, no matter what its constitution or charter, as the League of Nations was powerless and as the UN Security Council is powerless in the face of such disagreement. For instance, a CSCE effort to intervene in support of the independence of the Baltic states or of Georgia, Armenia, or the Ukraine is unlikely to succeed if the Soviet Union opposes it. If the major member states do agree, CSCE could, in part through a crisis prevention mechanism, bring to bear the combined political and economic influence of all member states to deal with frictions among the small states, as in southeastern Europe, the most plausible source of serious problems The chances are that such pressures will be successful in keeping the peace or in containing and limiting conflict if it docs break out Yet early experience showed that insistence on consensus frustrated most efforts by CSCE member states to object to domestic acts of repression by die Romanian, Yugoslav, and Soviet governments.

Its potential capability to reduce-interstate frictions alone makes a reinforced CSCE structure useful, but as one component among several of a multifaceted European structure. We have pointed out that the European Left is thinking in terms of movement toward a single all-encompassing European security structure, which would integrate the forces of the rival alliances on the model of the European Community. For the foreseeable future, a more realistic vision is of a new European security system composed of several mutually supporting elements: the European Community in the political and economic field; an on-going NATO: some bilateral treaties of assurance between a united Germany and its eastern neighbors, possibly with parallel treaties between the Soviet Union and the Eastern European states; sub-regional groupings; and an emerging CSCE structure.

This distinction between a multicomponent European security structure and a single overarching institution does not merely concern a transitional period of whatever time it takes to move from one to the other. Instead, a more basic question is involved: whether the Soviet Union will become a fully integrated member of the EC and of all European institutions to the same extent as current members. Theoretically, a stable and economically effective democratic system might be achieved in the USSR or its successor states after some decades, and a separate NATO or a successor European Defense Union will no longer be needed. Even so, given considerations of size alone, it is difficult to envisage the complete amalgamation or integration of this new Soviet Union or of a still very large Russia in a flourishing European Community. Some form of integration and mutual intertwining is the answer. However, a broader, more robust framework going beyond the EC and the CSCE may well be needed to integrate the Soviet Union, a framework that would include Japan and would cover the entire territory of the USSR and of Canada and the United Stales.

But that issue is for the distant future. Today's task, as well as the task of the coming decade, is to establish a more modest but nonetheless effective CSCE structure, with the full and active participation of the USSR and of the United States and Canada, as part of a new European security system whose main task will be to create confidence for the still more important work of political and economic integration. Such a CSCE structure will probably come into being by the mid-1990s.

9.5 NATO

A continuing NATO alliance is essential to the successful operation of the European security system, including CSCE security institutions Following negotiated East–West cuts in a continuation of the CFE talks that go beyond those anticipated in a first CFE agreement, and for the period of a decade or more, we expect that the U.S. contribution to NATO might comprise perhaps 120,000 personnel stationed in Europe, with roughly 70.000 in Germany—an army corps plus three to five air force wings. This is roughly the force left in Europe after the shift of U.S. forces to the Persian Gulf. A smaller two-division force, perhaps organized with European divisions into a multinational army corps, is also possible In either mode, these U.S. forces would be backed by air force reinforcement units based in the United States and by ground force reinforcements with heavy equipment propositioned in Europe.

In the new situation, NATO will no longer have a single encompassing mission to prevent a Warsaw Pact attack on Western Europe. Instead, it will have a series of missions.

- To plan defense strategy and force structuring of member states.
- To coordinate their arms control activities.
- To coordinate and evaluate the success of the policies and activities of member states toward the USSR, especially those aimed at the democratization of the Soviet system and of the other states now members of the Warsaw Pact, a key function.
- To provide assurance of the successful integration of a united Germany into the European system.
- To provide reassurance to the Eastern European states of their continued independence vis-à-vis the USSR—with or without formal guarantees from NATO countries; a changed Soviet leadership would know in advance that aggressive actions toward Eastern European states could ultimately involve NATO.

- To act as a barrier against spread or escalation of possible conflict originating in southeastern Europe.
- To support the CSCE, coordinating at least loosely the policy of NATO members toward the CSCE projects in the security field and by its existence as a force lending additional weight to CSCE decisions on security issues.
- To coordinate the defenses of member states against the contingencies of adverse developments in the USSR itself described above, providing reasons for any Soviet government to believe that military adventures would not be profitable.

In the long term, to secure its own evolution into a European Defense Community, the military arm of interested members of the EC, those member states that are prepared to engage in such an effort.

Because for the immediate future the European states will probably be on their best behavior with only limited friction among them, the last function noted above may be NATO's most important one. Presumably, the European Defense Union would include all the current members of the EC, except perhaps neutral Ireland. Currently, there is heavy debate in EC capitals as to whether the community should be 'deepened' to intensify its supranational character or widened geographically to include new member states. If the first view prevails, at least for a time, steps toward coordination of the foreign and security policies of member states will proceed more rapidly. If the member states decide instead to place more emphasis on broadening in order to envelop more fully the states of Eastern Europe, the community could expand to include members from the European Free Trade Area, such as Sweden and Finland But some of those states arc militarily neutral and some, such as Switzerland, may wish to remain so. In that event, it is probable that a European Defense Union—in the early stages, this role will probably be played by the Western European Union (WEU)—will have fewer members than the European Community and would remain separate from the EC for some time to come, formally joining with it at a later date. And until European influence in NATO has become visibly dominant. NATO will be unsuitable as a vehicle for coordinating military activities of European states outside Europe Europeans recognize the need for American leadership in the defense of Europe but are unwilling to accept it automatically in Third World endeavors. For this purpose, coordination through the Western European Union (WEU) seems more plausible.

If the new European security system has taken effective form and has helped to maintain the peace in Europe, in the later stages of NATO's transition to a European Defense Union, U.S. combat formations would be withdrawn, with the possible exception of some air force units armed with nuclear weapons. If developments in Europe permit the withdrawal of these air force units after a further decade or so. The relatively few remaining U.S. personnel would man logistics bases in Western Europe, perhaps in France, Italy, the UK. and also

Germany, to support units stationed in the United States that would remain committed to the defense of Europe.

The aim of the suggested deliberate evolution of NATO toward a European Defense Union is to provide a smooth transition that leaves intact both the North Atlantic Treaty and a still useful transatlantic military relationship, with a continuing habit of cooperation between a European military grouping and the United States. The long-range American security interest in Europe is not to preserve NATO as an institution or to assure the continuing presence of U.S. forces in Europe as a device for assuring American political influence. The long-term U.S. security interest is to assure a free and democratic Europe with which the United States can cooperate on a worldwide political, economic, and, if need be, military basis. Put negatively, it is to prevent decisive influence over the immense human and material resources of Europe by a hostile power, the grounds for U.S. involvement in both world wars and in the cold war.

Consequently, the long-term U. S. aim should be to ensure that the European Community, or some other Western European grouping, is effectively autonomous in defense. If this aim is to be achieved, a deliberate American policy of promoting it through promoting the evolution of NATO will be needed, it is doubtful that the aim can be achieved without deliberate U.S. support. Without a deliberate policy of promoting the Europeanization of NATO, there is a possibility that at some future point after a period of continued peace. American forces may be withdrawn from Europe and NATO dissolved, without a European successor organization being ready to replace it as an element of stability, among other things acting to counterbalance Soviet military power. At present, frictions with the Soviet Union leading to open conflict in Europe appear improbable. Yet the absence of coordinated military forces on the continent to counterbalance those of the Soviet Union could tempt a future Soviet leadership to increase its influence over Western Europe to a degree that could restrict the autonomy of the EC.

True, even after a further force build-down, most of the current member states of the EC will maintain sizable national armed forces of their own But unless these forces continue coordinated and a common defense purpose emerges, the EC will be an economic giant but a military dwarf, remaining under the military shadow of the Soviet Union and of the United States, and unable to play its full role in global affairs. The latter situation would create a military power vacuum of the traditional type.

To summarize, from the viewpoint of common European and U.S. interests, American military forces should remain in Europe until four developments take place: real progress toward a proven system of parliamentary democracy in the Soviet Union and Eastern Europe as well; the successful integration of a democratic Germany into the EC; the evolution of NATO into an integrated European military force that can counterbalance Soviet military power and make the EC autonomous in defense; and the emergence of a CSCE security system that can give practical expression to the determination of the European states to keep the peace.

This may be a conservative vision of the future of Europe in this era of good feeling generated by positive changes in Soviet policy. But it is desirable and probable that developments will go in this direction once Western governments turn from the necessary task of emphasizing the peaceful intentions of NATO in the context of gaining Soviet agreement to NATO membership of a united Germany to defining their own changed security needs At the same time, combining a NATO/Soviet Union force balance with effective arms control can obviate the baneful competitive and escalatory developments to which balance of power concepts have in the past led in Europe.

Achieving this outcome, like the continuation of NATO itself, is contingent on the capacity of Western governments to convince legislatures and public opinion of the ongoing need to devote adequate resources to military defense. Despite the strong and understandable feelings of some Europeans that NATO military forces will no longer be necessary to safeguard their security, it seems on balance-plausible that NATO governments will succeed in obtaining continuing majority support from their publics for this combination of smaller but effective armed forces and arms control.

9.6 Security Aspects of German Unification

During 1990 security aspects of the German unity issue were discussed in a number of forums, including the 2 + 4 Talks and in still more important bilateral German-Soviet discussions. Many of the points agreed will become integral parts of the emerging European security system. These include the central issue of NATO membership of united Germany, accepted by Gorbachev in his talks with Chancellor Helmut Kohl in mid-July 1990. A move that opened the way for unification under Christian Democratic leadership, confirmed in the first post-war all-German elections held in December 1990.

One major reason for German membership in NATO is, of course, the reassurance that the integration of the German military potential will give to apprehensive European publics. It is similar to the reassurance provided in the 1950s through integration into NATO of the newly formed armed forces of Federal Germany.

An equally important reason for German membership in NATO is to preserve the alliance both as guarantor of the new European peace order and as the nucleus of a future Western European defense grouping. Without a united Germany's membership in NATO, it would not be possible either to preserve the organization as a militarily effective force to meet short-term contingencies, or to develop it over time into a defense grouping to counterbalance Soviet military power. Without German membership in NATO and the possibility of continued deployment of U S forces on the territory of a German ally, it also might not be possible to retain in Europe U.S. combat formations and a visible link to the American nuclear deterrent, both valuable for a lengthy transition period.

The 2 + 4 Talks and the important bilateral German-Soviet talks have also had the aim of reassuring the Soviet Union and the many Europeans concerned over the implications of German unity that Germany's unification will not pose a challenge to the security of other states.

Several useful measures of mutual reassurance, among them a proposed non-aggression pact, a shift of NATO strategy, and institutionalization of the CSCE, were endorsed by the NATO summit in early July 1990. Then, after an active round of consultations in the spring of 1990 involving Gorbachev, Kohl, Bush, Baker, Genscher, and Shevardnadze, the problem of NATO membership was resolved in the Kohl–Gorbachev meetings in the USSR in mid-July 1990.

A number of important restrictions on German, Soviet, and NATO military capability were agreed in the July meetings. They are: (1) a united Germany will not possess or produce chemical, biological, or nuclear weapons, (2) the active-duty personnel of German armed forces will decrease to 370,000 within the next 3 or 4 years and remain no greater than that limit; this restriction was incorporated in the 2 + 4 Talks, (3) all Soviet forces will be withdrawn from GDR territory within the same period, and (4) neither foreign forces nor nuclear weapons will be deployed on the territory of the former GDR. These provisions, together with the guarantee of postwar borders in the 2 + 4 Talks and the extension of the security guarantees of the North Atlantic Treaty to the territory of the former GDR at German unification, will also be major components of the new European security system.

In my view, these measures might usefully be augmented by further agreements to the effect that: (1) no additional members of the Warsaw Treaty Organization other than the GDR would be permitted to join NATO, (2) neither the NATO-assigned forces of the united Germany nor those of other NATO stales would move into the territory of other present members of the WTO in addition to that of the GDR. and (3) except as part of multilateral agreements under CFE or CSCE, the Soviet Union would commit itself not to move its forces into the territory of states from which its forces have been withdrawn or that had no Soviet forces. (Article IV. 5 of the CFE Treaty prohibits the stationing of foreign forces without the consent of the host state). To fill out the list of components of the new European security system, these measures could be supplemented with overlapping Western or German and Soviet guarantees for the security of the Eastern European states and by mutual arrangements for thin-out zones along borders or by other measures of non-offensive defense and perhaps by sub-regional groupings in Central and Eastern Europe.

A further answer to the question of whether this new European security system will succeed in keeping the peace will be given by the long-term out-come of negotiations on reducing conventional forces in Europe, not only the CFE Treaty agreed in 1990, but also its successor agreements We turn next to an assessment of these CFE negotiations.

9.7 The CFE Talks

In responding to Mikhail Gorbachev's proposal of April 1986 for a new round of force reduction talks to replace the still convening but unproductive MBFR negotiations, Western leaders decided to make a very stiff test of Soviet leader ship's willingness to make real progress in conventional arms control. They proposed total elimination of the Pact's huge superiority in tanks, armored troop carriers, and artillery through reduction to a new equal level slightly below that of NATO's current holdings of these armaments.

To the surprise of most in the West, the Soviet Union moved rapidly to accept these terms. In December 1988, Gorbachev announced a sizable unilateral withdrawal of Soviet forces from Eastern Europe, including six tank divisions. At the opening session of the new negotiations on Conventional Forces in Europe (CFE) in Vienna in March 1989, Soviet Foreign Minister Shevardnadze accepted the Western concept of reduction to a new equal level below that of NATO's current holdings of the major armaments proposed for reductions.

There was further rapid progress. In May 1989, George Bush proposed add-ing to the reduction of tanks, troop carriers, and artillery, the reduction of combat helicopters and aircraft and also U.S. and Soviet ground and air force personnel deployed in the Central European sub-region of the Atlantic-to-Urals area to a level of 275,000 men each. The Soviets had proposed reduction of aircraft, helicopters, and of active-duty military personnel, but NATO had at first opposed this. In January 1990, the U.S. president proposed still larger reductions of U.S. and Soviet military personnel in Central Europe to a level of 195.000 for each country, with a ceiling of 30,000 on U S personnel stationed in Europe. Within a month, at the Open Skies meeting in Ottawa, Shevardnadze agreed to this proposal. However, as the CFE negotiations went into their end phase, the Soviet Union agreed in the context of talks on German unity to withdraw all its forces from East Germany within two to three years. Consequently, CFE limits on U.S. and Soviet personnel were dropped in return for a general pledge to negotiate on manpower reductions in CFE follow-on negotiations, The only specific personnel limit in connection with CFE I was the 370.000-man limn on German military personnel.

Conclusion of the CFE treaty means, according to NATO figures somewhat adjusted after a discussion of definitions, reduction of over 32.000 WTO tanks as compared to 3500 NATO tanks; over 25.000 WTO artillery pieces to 1,200 NATO pieces; about 4,000 Pact armored vehicles to 1,000 NATO armored vehicles; 1,600 Pact helicopters compared to 335 NATO helicopters; and according to revised NATO proposals, about 8,000 Pact combat aircraft in return for reducing 1,000 NATO aircraft. In most cases, these Pact reductions arc on a scale of 10–1 in relation to NATO reductions; the latter will be only 10–15 % from existing levels and, in some cases, far less.

The CFE treaty provides that reduced armaments be destroyed. Objections by the Soviet Union and Eastern European states over the costs of armament destruction made evident that the prescribed procedures were too complicated.

This feature of the treaty was cited by the USSR as partial justification for moving in 1990 a very large amount of reducible armaments cast of the Ural Mountains outside the area of application of the CFE treaty. Moreover, both sides are destroying older equipment in the reduction area. The remaining forces consequently will have modem equipment but there will be fewer forces. Not only the amount but the nature of these armament reductions is significant Tanks, armored troop carriers, artillery and attack helicopters, and aircraft arc penetration or outreach weapons essential for attack there is no substitute for them for the offense, although there are substitutes for them in the defense.

Other features of a CFE agreement are important. A ceiling is placed on bridging attached to armored vehicles. The reduction area is divided into subzones in order to control density and to prevent force concentration in the forward areas. The agreement contains provision for on-site ground inspection. Aerial verification is to be covered in CFE IA follow-on talks, which began in February 1991. The main consequence of a CFE agreement along these lines, plus withdrawal of Soviet forces from Eastern Europe, would be the end of the possibility of overwhelming surprise attack smashing through NATO's geo-graphically shallow defenses, which has been the main fear of Western governments for the past 40 years.

9.7.1 End Phase of the CFE Talks

In the summer of 1990, in contrast with the rapidity of early developments, the CFE negotiations had slowed down, although only temporarily The immediate cause was that the steady stream of Soviet concessions, which had provided the motive power for earlier rapid progress, including in other East–West arms control negotiations, had dried up The Bush-Gorbachev summit in Washington in May 1990 failed to make substantive process on CFE. The basic reason for the slowdown was the apprehensive reaction of Soviet leaders to the collapse of their position in Eastern Europe, which in practice almost reversed die balance of power in Europe. It left the Soviet Union on its own facing a NATO that would probably be augmented in potential strength by membership of a unified larger Germany.

As regards aircraft reductions, the main substantive area of East–West disagreement in CFE during the first half of 1990, the positions of both sides were close numerically. Both were arguing for a limit of 6,000–7,000 aircraft per alliance. But there was disagreement over definitions. The Soviets wanted to exclude from the reduction count about 1,500 armed and combat-capable two-seater training aircraft and their entire fleet of land-based bombers assigned to the navy, which they said they would not include in reductions or numerical ceilings unless NATO countries were willing to put a ceiling on their own carrier-based naval aircraft.

U.S. and Soviet opinion moved increasingly during 1990 toward postponing reduction of aircraft to a second phase of CFE talks. Only a few European countries were arguing for a further strenuous effort to include aircraft reductions in a first CFE agreement. In substance, they were right to do so. CFE participants have agreed that combat aircraft, with their speed, flexibility, and range, are a key weapon both for short preparation attack and sustained offensives. Aircraft capable of ground attack will become even more important once ground forces are reduced; these cuts will lead to increased emphasis on mobility. At the end of September 1990, agreement was reached on the main open issue, treatment of land-based Soviet bombers assigned to the navy. The Soviet Union agreed to accept a no-increase cap on these aircraft in return for a similar limitation on NATO's lower number of land-based naval aircraft.

The May 1990 Washington meeting of Bush and Gorbachev did result in an important agreement on timing of CFE. Bush agreed to participate in the summit meeting of CSCE states that Gorbachev had proposed, scheduled for mid-November 1990 in Paris, whereas Gorbachev agreed to the U.S. view that conclusion of a first CFE agreement was a precondition for holding such a CSCE summit. This compromise provided a mutual deadline to accelerate the CFE talks and virtually assured their rapid completion once the issue of Ger-man membership in NATO was resolved.

The breakthrough on the German issue came in the bilateral German-Soviet talks in the USSR in mid-July 1990. The German-Soviet agreement, supplemented by agreement in the 2 + 4 Talks the following day on the way in which a united Germany would confirm the present German-Polish border, meant in practice that a first CFE agreement would be concluded by early November. It also meant that as soon as a first CFE agreement was signed, follow-on talks called CFE IA would be convened on how to limit German military personnel and those of other negotiation participants.

The exchange of numerical data by participating governments when the CFE treaty was signed resulted in two disputes concerning Soviet forces. NATO governments claimed the Soviets had underreported their treaty limited equipment (TLEs) in the Atlantic to the Urals area. The Soviet government claimed that the treaty-limited arms of three motorized rifle divisions that it had just subordinated to the Soviet navy as coastal defense units did not fall under the limits of the CFE treaty, whose agreed terms of reference excluded naval forces. Differences over the first issue were narrowed after further discussion. Secretary Baker made clear that the treaty would not be submitted for ratification unless both issues were resolved.

Before dealing further with the possible content of follow-on CFE negotiations, we should review some of the probable shortcomings of the first CFE agreement because it would be in the Western interest to tackle them in any continuation of the CFE talks. It should be kept in mind that, in comparison to the positive achievements of a CFE agreement, these gaps are secondary. Moreover, they are correctable, because the negotiating process will not stop with the conclusion of a first CFE treaty.

9.7.2 Tasks for Future CFE Talks

A first CFE agreement is the beginning of the build-down *of* the military confrontation in Europe, not its conclusion. The forces of both sides will still be huge. The CFE treaty docs not contain effective limits on military personnel.

The absence from the treaty of specific limits on Soviet active duty ground and air force personnel stationed on Soviet territory west of the Urals creates an important potential loophole in CFE. If left uncorrected this omission would permit the Soviet Union on its own territory (as well as Western participants other than Germany) to increase the personnel of ground and air force units remaining after agreed armament reductions are carried out, increasing their effectiveness in the event of conflict. For example, CFE I signatories could increase the infantry or combat engineer components of armored divisions; establish additional divisions with fewer tanks and artillery, but more non-limited armaments; convert reserve divisions to active duty divisions by adding active duty personnel; establish a wide range of cadre units using limited numbers of active duty personnel, to be filled up on signal by reserve personnel; or build up the ground-based infrastructure of permitted aircraft, providing a higher sortie rate in the event of conflict. In other words, personnel reductions and specific personnel limits are a necessary pan of force reduction agreements in addition to reduction of armaments.

The NATO countries will seek to deal with these shortcomings in CFE IA through declarations by CFE governments that they will not increase their manpower. But with the exception of the 2 + 4 treaty commitment by Germany, these declarations will not be contractual, verified limits. Nor will they necessarily involve reductions this position should be changed.

Influenced by the negative experience of the MBFR negotiations, which were focused on personnel reductions, the NATO states earlier decided to omit man power from agreed cuts because of the verification difficulties they foresaw. Given the Western proposals for reducing United States, Soviet, and German personnel, and stabilizing measures limiting the size of out-of-garrison activities that abstention was—correctly—overcome in principle. If ceilings are focused on personnel in organized military units, with full exchange of data on personnel, armaments, and units, and with acceptance of intrusive verification, manpower reductions and limits of manpower can be adequately verified.

On another point, the terms of reference of 'mandate' of the CFE talks agreed by all 23 participants in January 1989 called for "the elimination, as a nutter of priority, of the capability for launching surprise attack and for initiating large-scale offensive action." The objective of eliminating surprise attack is being met adequately in the first CFE agreement. But a first CFE agreement will not do enough to meet the second agreed objective, eliminating capability for sustained large-scale offensives.

One major reason for this shortcoming is the absence of provision for specific ceilings on military personnel and on the number of combat units in the Atlantic-to-the-Urals area. Moreover, the CFE I agreement will contain no effective measures impeding rapid forward movement of forces. It will contain no restrictions on mobile equipment that accompanies the forces, like armored antiaircraft weapons or field hospitals, or on logistic components, like pipe-laying equipment, tank munitions and fuel transport units, or bridging equipment other than that attached to armored vehicles. And it will not restrict training of reserve personnel for mobilization.

Owing to pressures from arms-selling NATO member states, plus the refusal of the United States to extend measures of this kind to its own home territory, CFE will also do nothing to limit the production inside the Atlantic-to-Urals reduction area of armaments of the type reduced by agreement. These arms could be stored at production plants for later use by oversized personnel complements of organized units. Instead, newly produced arms should be made more observable through storing them in designated locations open to air or ground observation. Tb achieve agreement on this by European countries and the USSR, it may be necessary to place arms production in USSR territory outside the Atlantic-to-Urals reduction area and in the United States under the same modest reporting requirements as inside the Atlantic-to-Urals area.

Modernization of permitted anus could be an even greater problem than at present under conditions of arms reductions where the traditional counteraction to introduction of new arms, increase of other armaments, may be blocked by treaty limits on armaments of that kind. More needs to be done in this field, but a confidence-building measure approved at the Paris summit would require a year's advance notification of deployment of newly developed weapons. A further confidence-building measure calls for exchange of defense budgets and their discussion If the requirement for budget information can be developed to call for more detail than is now the case, this measure could make it possible to raise questions about major modernization projects years in advance of their actual production and deployment.

Verification measures for CFE will be stringent, but they will omit permanent observers at major military installations, traffic choke points, or exit-entry points into the Atlantic-to-Urals reduction area, as well as traveling go-anywhere patrols; all of these measures would be useful. These omissions should be tackled in continuing CFE negotiations.

Another negative by-product of CFE I should also be tackled in continuing negotiations. This is the fact that, unless some further constraints are applied, military planners on both sides, no longer having sufficient heavy conventional armaments to pursue a linear defense along their entire borders, and with NATO and Soviet forces separated in Central Europe by the intervening territory of Poland and Czechoslovakia, will probably emphasize rapidly moving armored units and air-delivered forces. Birth sides will be equipped with still very large and modernized holdings of weapons of this type, which are, as noted, essential for the offense. Both sides are likely to employ a strategy of maneuver warfare and a meeting engagement.

Even with peaceful intent, with both sides similarly equipped with a high number of heavy armaments indispensable for attack and with preemptive speed an important value, military stability could suffer. If either side fears the other is about to attack, the advantage will go to the side that strikes first and gets farthest and fastest with its armored columns. There arc three eures for this situation: further deep cuts in offense-orientated armaments; restrictions on logistics essential to rapid movement of forces; and a defense-oriented posture in border regions, with mobile elements pulled to the rear. Agreement on measures of this kind arc needed if the two alliances are to make definitive progress toward the objective of defense dominance, which both have accepted as the long-term objective of the Vienna talks.

9.7.3 Continuing CFE

The Western objective in continuations of the CFE talks should not only be to make good these shortcomings of CFE I. The overall Western objective should be to cut NATO forces to a lower equal plateau with the USSR, providing force levels that NATO countries can hold with less strain and cost over the next decades while they observe long-term developments in the USSR. At the same time, NATO's objective should be. while the opportunity continues, to use the willingness of the Gorbachev government to negotiate further build-down of the East–West military confrontation in Europe to create additional stability between remaining forces and to build in further barriers of delay against possible decisions of a less cooperative Soviet government to build up Soviet forces.

NATO governments should also use the CFE process to save money. With reductions of military personnel and armaments of 40% from CFE I levels, European NATO countries should be able to save at least 30% of their current combined defense budgets of $150 billion a year. If U.S. forces in Europe were reduced to under 150.000 from the current level of over 300.000, and the withdrawn forces disbanded after solution of the Gulf crisis, this should also bring savings of up to $30 billion a year. Consequently, the next set of CFE talks should be conceived not only as narrowly focused on dealing with the issue of German military manpower, but as far broader in scope.

But neither the Soviet Union nor the Eastern European states nor European neutral states wish to continue the CFE talks on the current alliance-to-alliance basis. The Eastern European states do not wish to be obliged to seek common Warsaw Pact positions, and the Soviet Union docs not want to undertake the difficult task of trying to achieve unified positions or to accept sub-limits that restrict it to only 60–70% of the total of personnel and arms accorded to the Warsaw Pact as a whole, which therefore make it numerically inferior to NATO.

To meet these difficulties, a changed negotiating concept for continuing the CFE talks will be needed. The CFE concept of reducing to equal levels should be retained but applied in a different way. In this changed approach, parity at lower levels would be sought between the forces of all NATO states or of a part of NATO states, and all or part of the territory and forces of the Soviet Union west of the Urals. If all NATO states do not wish to reduce them, Britain, France, the United States, and the Benelux countries can join Federal Germany, already committed to large personnel reductions, in reducing their forces deployed in Germany to a lower equal level with Soviet forces in the Baltic, Byelorussian, Carpathian, and Kiev military districts (see the author's study referred to at the beginning of this chapter). The Eastern European states and those NATO states not participating in the organized reductions would participate in the continued negotiation and would commit themselves either to make individual reductions or to accept enduring upper limits on their force levels. This approach would combine reduction of NATO and Soviet military personnel with further reduction of the same armaments reduced in CFE I—tanks, artillery. Armored personnel carriers, combat helicopters, and aircraft. The objective of a CFF II of this type should be cuts of between 20 and 40% in both heavy armaments and active duty personnel from the NATO's post-CFE I levels, with Soviet forces coming down to NATO's new level, either for the wide Atlantic-to-Urals area or in an enlarged control zone. To the extent possible, reductions should be by units, with no-increase limits on remaining units of the type reduced.

9.8 Some Conclusions

As we have seen, the unification of Germany, the conclusion of a CFE treaty, and a CSCE summit that approved additional confidence-building measures and took the first steps toward establishing permanent CSCE security institutions lor Europe were the major prospective developments of 1990 for European security. They are likely to be followed within the next three to four years by a further CFE agreement covering at least personnel reductions; a further agreement on confidence-building measures, an agreement on short-range nuclear forces, and by further development of CSCE institutions. It is likely, too, that in that period the NATO alliance would cope with its problems of reforming and transforming itself to meet the new situation.

The European security system that results from the combination of these components is likely to have at its disposal organization and measures that arc adequate to deal successfully with the main objective of the system—preventing interstate conflict in Europe. The new security system is likely to be less successful in efforts to avoid future conflicts through intervening in domestic frictions. However, whether the new European security system will actually successfully cope with these challenges will depend less on its institutions and other components than on the willingness of its member governments to make full use of the assets of the security system to meet challenges to the security of Europe. As long as the memories of World War II and of the Nazi and Soviet dictatorships remain strong, the governments seem likely to make the right decisions.[2]

[2] *Editors' Note* The disputes between the United States and the Soviet Union regarding Soviet forces and CFE treaty limits (p. 63) were resolved in Lisbon where, in late May 1991, Foreign Minister Bessmertnykh and Secretary Baker had intensive talks. The results cased the way for the ratification of the CFE treaty in late 1991 or early 1992. (Sec "Conventional Arms in Europe, 1990 and 1991," *Arms Control Reporter,* Section 407, May 1991).

Jonathan Dean visiting the People's Congress in Beijing during a conference on "Security, Arms Control and Defence Restructuring in East Asia", co-organized by COPRI, UNIDIR and the Chinese People's Association for Peace and Disarmament in Beijing in December 1996. *Source* Personal photo collection of the Dean family

Chapter 10
Arms Reduction Blueprint (1991)

In the last 2 years the threat of conventional and nuclear war in Europe leading to worldwide nuclear war has receded far into the distance.[1] However, the possibility of nuclear war from Soviet territory or elsewhere in the world has not ended. In fact, the collapse of central power in the Soviet Union has been accompanied by a serious decrease in Soviet accountability for nuclear weapons, a very negative development.

Despite official statements to the contrary, Soviet strategic nuclear forces—and this is also largely true of US strategic forces—are configured for launch on warning. As long as both states have strategic nuclear forces large enough to launch a surprise attack with a fairly good chance of success, both will be obliged to configure the bulk of their strategic forces for rapid response. A risk of accidental launch will therefore remain built into the system. If, in time, there is a change of government in the Russian and other Republics—and a trend to more autocratic rule seems probable—there will be increased possibility of political friction inside the Soviet Union and with the outside world, and these risks will mount.

The current situation with regard to custody over nuclear weapons, both strategic and tactical, is not reassuring. President Gorbachev has recently signed the

[1] This text was presented as a keynote speech on Friday, 26 October 1991 during the Seventh AFES-PRESS Conference in Mosbach, Germany and published as: "Arms Reduction Blueprint," in: Hans Günter Brauch (Ed.): *Controlling Military Research, Development and Exports of Dual Use Technologies as a Problem of Disarmament and Arms Control Policy in the 1990s—The Results of the Seventh AFES-PRESS Conference, Abstracts and Discussions*, AFES-PRESS Report No. 45 (Mosbach: AFES-PRESS, 1992): 75–81.

H. G. Brauch and T. Grimwood (eds.), *Jonathan Dean*,
SpringerBriefs on Pioneers in Science and Practice 19,
DOI: 10.1007/978-3-319-06662-2_10, © The Author(s) 2014

fifth draft treaty regulating the relationship among the Republics that once constituted the Soviet Union, which appears to give a central organisation day-to-day responsibility for command and control and of custody over nuclear weapons, with decision on use to be shared among the Republics. That would be good. But, whatever agreement is reached finally on paper, the relationship among the Republics is likely to remain turbulent for years to come, and this nuclear issue with it. Right now, some of President Yeltsin's influential advisers are arguing that there should be no central government whatever in the former Soviet Union and that issues of custody over nuclear weapons, like other issues, should be settled between the Russian Federation and other Republics.

The Ukraine and Kazakhstan have said they want no nuclear weapons stationed on their territory, but have also indicated that they do not want the Russian Republic to have monopoly control over such weapons and that they will not relinquish their custody until other aspects of their relationship with Russia are settled. The Ukraine is also saying it wants itself to destroy the nuclear weapons now on its territory in order to cut the overall number and to decrease the Russian monopoly. The Byelorussian Defence Minister, whose government is composed of unreconstructed Communists, has justified retention of nuclear weapons in Byelorussia by the argument that NATO still has nuclear weapons.

Custody over nuclear weapons on the territory of the Republics is now exercised by Union armed forces, but the Republics say that only their own armed forces should be stationed on their territory, and the Ukraine has begun to 'Ukrainise' units of the central armed forces, passing a law transferring to Ukrainian sovereignty all military forces on its territory, conventional or nuclear, including the Soviet Black Sea Fleet.

There are further proliferation risks: the Russian Republic has about 10,000 tactical nuclear weapons on its territory, but 5,000 more are scattered in the other Republics. It appears that some of those pulled back from Eastern Europe have been stored in the Kiev area in the Ukraine. There is considerable risk that they could be taken over by terrorists for exploitation, perhaps as blackmail to involve outside powers, or to be sold outside the country. There are also many trained and employed nuclear physicists scattered in nuclear research labs and energy-producing reactors among the Soviet Republics, some of which may in time seek to develop their own nuclear weapons.

The more immediate risk is of hard currency sale of nuclear expertise and components to Third World countries, like Pakistan, North Korea, and Libya. In 1990–1991, the Soviet central government sold nuclear power plants to Pakistan and to India—two of the mutually hostile undeclared nuclear powers, non-signatories of the Non-Proliferation Treaty—without requiring that they accept full safeguards on all their nuclear plants. The majority of Soviet uranium-processing plants are in the Asian Republics, Kirghizstan and Tadzhikistan, far from the weakening control of Moscow, creating the possibility of hard currency sales to Third World countries like close-by Pakistan and India.

10.1 Control and Accountability

It seems clear that, whatever the long-term relationship among the Soviet Republics, Western countries have a strong security interest in central command and control and accountability over existing nuclear weapons and weapons-producing installations, whether exercised by the central government alone or in collaboration with the Russian Republic. We cannot object if the Republics want co-decision over possible use of nuclear weapons, but there should be no mixed responsibility for command or for day-to-day control over existing weapons.

We also face the problem that if there are no controls whatever on the level of nuclear arms of the other *declared* nuclear powers, especially China, where similar risks of slipping into regional control over nuclear weapons could occur. North Korea is a dangerous puzzle, and there are also no controls on the weapons of the three *undeclared* nuclear states—Israel, India, and Pakistan.

The risks of development of nuclear weapons by non-nuclear states has increased with the wider availability of expertise and technical knowledge. This enabled Iraq to engage in three different methods of developing nuclear weapons, unknown to Western intelligence or to the International Atomic Energy Authority, capabilities which, according to IAEA inspectors, could have resulted in systematic production of nuclear missiles within 2 years.

In my view, President Bush's welcome nuclear initiative of 27 September is positive in all but three aspects: continued funding for the B-2 when we already have other aircraft which can carry out conventional bombing missions; the proposal for a limited nationwide missile defence; and a rather difficult position on the desirable objective of eliminating multiple warheads of MIRVs. Especially worthwhile elements of the President's initiative are:

- The radical decision to withdraw ground-based tactical weapons on a worldwide basis and to destroy the warheads. Destruction is a necessary and overdue step.
- The decision to withdraw and partially destroy tactical-range nuclear arms from surface ships and submarines.
- Actions to reduce the readiness status of US forces, including taking bombers off alert, and the proposal for co-operative measures with the Soviet Union to improve custody and control over nuclear weapons, their storage, transport, dismantling and destruction. This, the most urgent aspect of the Bush proposals, has been accepted by Gorbachev.

President Bush's proposal to withdraw MIRVs from land-based ICBMs was not new and has already been rejected by the Soviet Union several times. Gorbachev has questioned it again.

On a rough calculation, the US proposal would mean reducing only about 500 strategic MX warheads from the US post-START level of over 10,000 warheads, and about 2,000 from the post-START Soviet level of 8,000, leaving the US with nearly twice as many warheads as the Soviet Union. Even if this were agreed, there still would be too many warheads on both sides for real stability. In the interest of dispersal, the submarine-based MIRVs might be cut back sharply, but this could be costly in terms of immediate requirements for new submarines smaller than the Leviathans both countries have been building. An interim step would be to move toward a lower, more equal level of MIRVs.

The President cancelled research and development on the rail-based version of the MX missile on the mobile version of the small single warhead missile now in development, and on the air-mounted short-range attack missile, including the version planned for deployment in Europe. He did not mention the current programme for development of manoeuvrable warheads, MIRVs or earth-penetrating warheads.

10.2 Soviet Position

In his 5 October response, Gorbachev agreed to Bush's challenge of parallel unilateral actions: destroy all ground-based tactical nuclear weapons; remove all tactical-range nuclear weapons from surface ships and submarines, and destroy a portion of the nuclear warheads of anti-aircraft missiles; continue to keep Soviet bombers off alert; stop development of new mobile missiles; freeze the level of rail-mobile ICBMs and keep them in base; and remove some ICBMs from alert.

President Gorbachev also proposed a moratorium on nuclear testing; negotiation on a further post-START cut of 50% in strategic arms (apparently his answer to the Bush proposal to remove MIRVs from land-based missiles); destruction as well as withdrawal of all tactical naval weapons; and separate storage of aircraft-mounted nuclear bombs. Reversing the earlier Soviet opposition to missile defence, which has gradually been weakening, he agreed to discuss the possibility of developing a system of defences against nuclear attack, plus creation of a joint early-warning system with both space- and ground-based components.

We need to develop a concept of the long-term Western nuclear relationship with the Soviet Union and its successor states, and some indication of where Western nuclear weapons policy should be headed. Failure to articulate these long-term goals soon could mean the Western coalition missing the present opening for a general breakthrough on nuclear weapons.

There is also a time factor with regard to China. A drift toward regional control over China's nuclear weapons may set in fairly soon, as Deng and the older generation move off the political scene. If regional party leaders or military warlords gain control, it will be a good deal harder to introduce an agreed regime of limits and restrictions on these weapons.

10.3 Programme for Security

Consequently, there is a good case for using the present opening to move toward eliminating the potential for surprise attack from the Soviet Republics and for coming to grips with the proliferation problem. The Union of Concerned Scientists is proposing such a long-range programme, comprising four main elements:

First, negotiated reduction of US and Soviet nuclear warheads of all types to the level of 1,000–2,000, what we are calling a *reserve nuclear force*, to be used only to deter nuclear attack or to respond, should it occur. We are also urging a series of co-operative actions designed to make residual forces and command and control less vulnerable. The US has now proposed many of these actions, and Gorbachev has agreed to work on them. We define nuclear stability between East and West as the absence of capability for surprise attack. Measures to reduce alert and readiness levels of both sides are essential and must be combined with drastic cuts in warhead levels.

Secondly, further actions in both countries are needed to make these reductions effective, not only by dismantling warheads, but also by permanently closing down production of fissile materials, and all but one weapons production plant in each country (its output to be verified), a step-by-step close-down of nuclear testing, and placing all nuclear installations, civilian and military, under a combination of bilateral and IAEA controls. Cutting Soviet and US nuclear weapons to a minimum level not only reduces risks of accidental launch, but is necessary to trigger decisive moves toward effective control over nuclear weapons elsewhere.

The third major component of our proposal, therefore, is to bring the other declared nuclear powers—the UK, France and China—into the regime of agreed low limits. We propose a limit of 500–700 warheads for each of these countries as well as ending their weapons production and testing. We urge prompt action to bring these countries into a system of international limits not only to control their nuclear potential, but also to enlist their cooperation in establishing a non-proliferation regime with real teeth.

This new regime would include renewal of the Non-Proliferation Treaty with supplementary agreements mandating tougher IAEA inspection practices, and a new action role for the UN Security Council in controlling nuclear weapons. One suggestion is that the Security Council agree to act as the enforcement arm of the Non-Proliferation Treaty, ordering inspections of all suspicious sites in member state territories. Secondly, to deter use of nuclear weapons by any state or terrorist group, we urge that the five declared.

Gorbachev has proposed this action, but as an independent measure; what I have in mind would make such a decision far more operational.

10.4 No First Use Ban

In my view, these five countries should then support a Security Council resolution making the threat of no first use of nuclear weapons an offence against world peace, against which joint action would be taken under the UN Charter. Countries or terrorist groups with nuclear weapons would then know in advance that the Security Council would come after them with punitive action—what Saddam Hussein did not know when he invaded Kuwait.

A third proposed Security Council measure has to do directly with the *undeclared* nuclear powers. We are proposing a Security Council decision to the effect that the failure of any state to accept minimum controls on its national nuclear installations would be an unacceptable threat to international peace and security and would incur economic sanctions and other measures. "Minimum controls" would consist of a wide range of options, including the signature of the Non-Proliferation Treaty, acceptance of bilateral or multilateral inspection like that between Brazil and Argentina, or participation in a verifiable nuclear weapons free zone, like that proposed by Israel, Egypt, and other countries.

As a token of the Security Council's determination to act against those states which do not comply, we would add a further rather controversial option—kept open only for a specified grace period, perhaps 6 months—whereby existing undeclared nuclear states could declare themselves as nuclear weapons states, while also accepting the full range of limitations already agreed to by the declared nuclear powers, including limits on the size of their forces, close-down of fissile material production plants, and a test ban and international control of all their nuclear facilities.

I believe that co-operation among the five permanent members of the Security Council on controlling nuclear weapons will promote greater co-operation between them and other Security Council members on peace-making and peace-keeping. A beginning must be made to articulate the long-term goals and to set out toward them.

One important by-product of the collapse of central Soviet structure is that it eliminated the last vestiges of what has previously been the strongest argument against missile defences—that testing or deploying them could lead to uncontrolled expansion of Soviet strategic missile forces. The present argument against missile defences—that testing or deploying them could lead to uncontrolled expansion of Soviet strategic missile forces. The present divided Soviet Union could not organise or finance such an effort. But, despite Gorbachev's agreement to discuss the matter, there are still strong arguments against even a thin nation-wide missile defence system, and these also apply to Europe. Among those arguments are that a combination of tough non-proliferation measures, controls on sale of missile parts, and flexible conventional forces, is a more effective means of dealing with potential threats than missile defence. The effectiveness of missile defence is also questionable and cannot prevent nuclear weapons being smuggled into target countries as a suitcase bomb or in a rusty tanker.

Missile defence is expensive and would compete for scarce resources with conventional forces; and it would downgrade the effectiveness of the nuclear forces of the other declared nuclear powers. It would therefore be an obstacle in gaining their co-operation to accept limits on their own nuclear forces and collaboration in a more active UN non-proliferation policy.

Deployment of missile defence, together and continued possession of large nuclear forces, would give the US and Soviet Union a second level of privileged status compared to non-nuclear countries, thus impeding efforts to extend and improve the Non-Proliferation Treaty, which lapses in 1995.

10.5 Anglo-French Dimension

While the nuclear situation in the Soviet Union remains unclear and political turbulence continues, it seems sensible to retain US nuclear weapons in Europe to provide confidence. Continuation of a US role may also be helpful in convincing non-nuclear states in Europe—and in Asia—that their security interests do not require them to develop their own nuclear weapons.

It may also promote a move toward Franco-British nuclear co-operation. In the long term, a relationship between a completely denuclearised European Community and a Russian Republic or Republics with a continuing nuclear force, even one reduced and heavily controlled, would not be healthy. But it will be long before the European Community is prepared to manage its own nuclear force, and American participation may be helpful for that reason.

I believe too that moves by France and Britain toward greater co-operation in nuclear weapons will get far more public support if they are part of a larger enterprise in co-operation with the Soviet Republics toward more effective control over nuclear weapons everywhere. Franco-British nuclear all aspects of nuclear proliferation and to promote more effective peacekeeping.

Chapter 11
Constraining Technological Weapons Innovation in the Post-Cold War Environment (1992)

Abstract In the period following the fading away of Cold War military competition, the United States seems likely to be the main source of technological innovation in weapons. (The abstract was first published in: Hans Günter Brauch (Ed.): *Controlling Military Research, Development and Exports of Dual Use Technologies as a Problem of Disarmament and Arms Control Policy in the 1990s—The Results of the Seventh AFES-PRESS Conference, Abstracts and Discussions*, AFES-PRESS Report No. 45 (Mosbach: AFES-PRESS, 1992): 162. Permission was granted by AFES-PRESS on 6 March 2014). Despite continuing downward pressures, it will probably have the largest defense budget and the largest research budget. It is neither feasible nor desirable to abolish innovation in the abstract. It is feasible to slow or suspend the process as it concerns specific weapons to be introduced in specific contexts. In the post-Cold War period, cooperation among the member states of the Conference for Security and Cooperation in Europe, including the United States and the Soviet Union or its successor states, will provide a positive motive for caution in innovation, the desire to avoid developments which could undermine or imperil that cooperation. This consideration, plus the absence of a major challenger, may prove as powerful a motive for caution in innovation as the consideration, seen at its most effective in the US- Soviet agreement to limit defenses against ballistic missiles, that introduction of new weapons could upset the East–West force balance and trigger a new round of competitive force improvements. It is argued in this chapter that, as evidenced by the low effectiveness of the assessments of the potential impact of new weapons systems prepared by the Arms Control and Disarmament Agency in the United States, it is difficult for governments to abstain from innovation if their decisions remain wholly unilateral. Even so, it is suggested that impact statements would gain more weight on the decision-making process in the United States if they were prepared by the Office of Technology Assessment on behalf of the Congress. The Congress, by its political composition and traditional role in counterbalancing the Executive Branch, is likely to be more skeptical about the desirability of specific weapons innovations than the Arms Control and Disarmament Agency, which now prepares these assessments. The Agency is a part of the Executive Branch of the United States government. The possibility that

H. G. Brauch and T. Grimwood (eds.), *Jonathan Dean*,
SpringerBriefs on Pioneers in Science and Practice 19,
DOI: 10.1007/978-3-319-06662-2_11,

destabilizing innovations could be slowed or stopped would increase if the decision-making process became less unilateral, and if there were authorized channels for conveying and discussing the views and objections of outside governments to the innovating government. With that in mind, it is suggested first, that a consultative process to discuss innovations be established among the five nuclear powers and, second, that two *CSBMs* contained in the Vienna document of the *CSCE,* in their present form providing for preannouncement of deployment of certain new weapons systems and for discussion of defense budgets, be expanded to become more effective vehicles for discussion of military innovation among *CSCE* members. Discussion, 'jawing' as Churchill called it, will not stop technological innovation -nothing can do so in the absolute sense- but discussion can slow innovation down.

11.1 Introduction

Other chapters of this book describe an expanding array of cooperative international efforts dedicated to restricting the spread of military technology.[1] Merely to illustrate, these efforts include:

- Prohibition of development, production, or possession of weapons, as in the case of the Biological Weapons Convention, the US–Soviet Treaty on Intermediate Nuclear Forces, the Non-Proliferation Treaty as it applies to non-nuclear signatories, and the convention on chemical weapons now under negotiation in Geneva;
- Commitments not to possess or deploy specified weapons in specified areas, as in the treaties establishing nuclear-free zones in Latin America, the Antarctic, Outer Space, and Seabed Treaties;
- Commitments to constrain testing and development of specific weapons, as in the Limited Test Ban Treaty and the Treaty on the Limitation of Anti-Ballistic Missile Systems (*ABM* Treaty), and the treaty proposed by the then *USSR* against deployment of weapons to destroy satellites (*ASAT*), as well as the proposed Comprehensive Test Ban Treaty; and

[1] The text was published as: "Constraining Technological Weapons Innovation in the Post-Cold War Environment", in: Hans Günter Brauch, H. van der Graaf, John Grin and Wim Smit (Eds.): *Controlling the Development and Spread of Military Technology* (Amsterdam: VU University Press, 1992): 35–41. The VU University Press does not exist any longer and thus the co-editors granted permission to republish this text.

- A whole array of measures to restrict the sale and transfer of components for the production or construction of weapons, including nuclear components, missile components. Chemical weapons, specialized machinery, super-computers, and high-tech arms.

These efforts are impressive. But arrayed against them arc still more powerful forces driving the continuing proliferation of all types of weapons: the desire of governments to have more powerful armed forces, their willingness to pay high prices for them, and the desire of sellers to gain economic benefits from the world's largest business. Above all, as shown in the ongoing case of Iraq, weapons proliferation is sustained by a rising world level of technological and engineering capability and the spread of technological knowledge through publication and training of scientists and technicians. Taken by itself, the collapse of the Soviet Union will free the services of several thousand nuclear weapons technicians for potential service throughout the world.

Innovation, the invention of new weapons and the improvement of existing weapons, is an especially effective engine of weapons proliferation and of competitive arming between adversaries. Time after time, this competitive effect was at work in the *NATO-WTO* military confrontation. We are seeing it again in the interest shown by 'Third World' defence ministries in the advanced munitions that coalition forces used in the Gulf War against Iraq.

Over the centuries, weapons technology has advanced with accelerating rapidity, producing weapons of increasing destructiveness, range, and accuracy. It has been a long-held hope of people everywhere that ways could be found to slow this powerful advance (Smit 1990). The advent of nuclear weapons, the Cold War, and the East–West military confrontation which was both a product and an intensifier of the Cold War, heightened this interest. Soviet scientists and officials, who correctly saw the Soviet Union at a disadvantage in technological weapons confrontation with the United States, have frequently raised the issue of action to slow this development.

11.2 The Case of MIRV

Multiple nuclear warheads (multiple independently targetable re-entry vehicles or *MIRVs,* a package of two or more re-entry vehicles with nuclear warheads carried by a single ballistic missile which can be targeted to separate targets) arc a dramatic example of this competitive innovative process. No noteworthy protest was raised against the development of missiles of intercontinental range either in the United States or the Soviet Union, where they were first developed in order to counterbalance the forward deployment in Europe of American bombers equipped to deliver nuclear weapons. But protests did come from the Soviet Union and from the United States Congress against development and deployment of multiple warheads.

This innovation made it possible to multiply by ten the number of warheads carried by a single missile. Consequently, it markedly increased the feasibility and advantages of a surprise attack on the strategic forces of the adversary state. As predicted by those who sought negotiated prohibition of this innovation, deployment of *MIRVs* resulted in a very large increase of warheads deployed by both the United States and the Soviet Union. Now, 20 years later, after expenditures of billions of dollars and after decades of bilateral relations poisoned by mutual concern over the possibility of surprise attack, the United Stales and the Soviet Union (or, after the recent changes, the Commonwealth of Independent States) have begun to 'down-load' and reduce the number of these multiple warheads mounted on their launchers, and they will try in future negotiations to restrict them further. In his proposal of September 1991, President Bush proposed the ultimate elimination of land-based Soviet and American *MIRVs,* mounted on ground-based intercontinental ballistic missiles (the President specified only land-based *MIRVs,* where the Soviets have numerical superiority. Presumably, if the Soviets are to agree, some reduction of *MIRVs* from submarines equipped to launch strategic-range ballistic missiles, where the *US* has numerical superiority, will also have to be included).

But, once these innovations are developed and deployed by their originators, those with the largest military budgets, they are sold, transferred, stolen, or imitated by others. Today, the UK and France arc in the process of 'mirving' their submarine-launched nuclear missiles, the *UK* by use of missiles bought from the United States.

11.3 Can or Should Weapons Innovation be Prohibited?

Now that the Cold War has ended and the Soviet Union itself seems no longer to exist, the main motor for technological weapons innovation during the past 40 years, the intense political and military competition between the United States and the Soviet Union, has been eliminated. Defence budgets will decrease, more rapidly in the former Soviet Union than in the United States, and the amount of money devoted to military research will also be cut. Budget reduction is the single most important restriction on weapons innovation.

Is the issue of technological innovation now laid to rest as a consequence? Will military innovation stop as a result, and the world's armies satisfy themselves with freezing military technology at present levels? Surely not. As long as one ingenious man is thinking of weapons improvement, the possibility of innovation remains. As long as armed forces exist, it will be argued that they should be equipped with the best possible weapons. In the abstract, this argument is irrefutable; it cannot be argued convincingly that armed forces should deliberately be equipped with second-rate weapons. And the argument will always be made, again with some justice: "If we don't innovate, the adversary will." Always the element of competition plays a major role, most powerfully when the adversary is known and identified, but also when he is an invisible scientist working in the back rooms of the defence establishment of an unlikely country, like Iraq.

We should perhaps stand back for a moment and look at this topic in more general terms. The subject of this chapter is how to place conscious controls on the human characteristics—individual and societal—which have brought humankind from the use of the rock as a weapon to the nuclear missile, but also to the use of the same missiles to launch instruments and human beings into space for exploration. In other words, we are called upon to separate that which is probably inseparable, to separate technological innovation in weapons from other types of innovation and to control the first without limiting the second. Thus, prohibition of technological innovation in weapons would probably be equivalent to limitation of all technological innovation.

The ideal that all disputes, disputes between states and groups as well as between individuals, should be settled by peaceful means is a noble one, and it should be urged on world governments. At the same time, there is little prospect in the present stale of world affairs that organized violence or organized armed forces can be eliminated. Thus, the existence of military forces does not represent a negative, and it is probable that the future will see many cases in which the judicious use of military force can save lives. This was true of the coalition that intervened against Saddam Hussein in the Gulf War of 1990–1991; subsequent discoveries have shown that this aggressive dictator was moving toward secret production of nuclear weapons on a large scale. Constructive, limited use of military force from the outside might have prevented many deaths in the Yugoslav Civil War.

11.4 Restricting Weapons Innovation in the Post-Cold War Period

In future years, the United States seems likely to continue to play the role of chief source of military innovation for the world's armed forces. Its military budget will prob-ably remain the world's largest despite continuing downward pressure. American cultural values arc important here: a central American value is a firm general belief in the virtue of technological innovation, this value is rooted in the American historical experience and development of innovations like the cotton gin and the grain reaper. Derived from this basic value is a belief that, in military matters, effective technology is the best substitute for scarce and expensive manpower, and a belief that, faced with the choice between cutting personnel levels and cutting military research, the best hedge against future contingencies is to maximize research. The absence of the Cold War adversary, or of any adversary of equal size, will brake the momentum of technological innovation in the United States. But in a one-sided situation of this kind, it also will no longer be possible to argue that a specific weapon could destabilize the overall confrontation or lead to a new round of arms competition.

Experience also shows that discussion about limiting technological innovation is most effective when it is conducted in terms of specific weapons systems or in the context of a specific military relationship, rather than in the abstract. An example that this is so, thus far at least, is the ABM agreement to limit the development and deployment of defence systems against missile attacks. In the context of the US–Soviet confrontation, the most effective argument against all-out development of missile defence systems by the United States has been that they would affect the force balance and elicit an even greater Soviet deployment of offensive missiles. Owing to the collapse of the Soviet threat, this argument is losing its potency. A visible result of this loss was the progress made during 1991 in the United States Congress of legislation providing for a 'thin' nationwide defence against ballistic missiles.

The United States experience also appears to point to a further conclusion: efforts to limit technological innovation arc more effective when they involve external reaction or pressure from allies or adversaries, and especially when they involve some kind of agreement with another government or governments, rather than depending on unilateral abstention by the developing country. This point is well illustrated by the ineffectiveness of the Arms Control Impact Statements which the United States Congress has legislated should be produced each year by the Arms Control and Disarmament Agency (*ACDA*).

This legislation (contained in a 1976 amendment to the law establishing the Arms Control Agency) was motivated by the idea that an independent assessment of the potential impact of major new weapons systems on existing arms control agreements or on the force balance in the Cold War would be a valuable contribution to rational Congressional decisions on funding of new weapons systems. This is a good idea in the abstract. But actual practice has shown it to be ineffective. Instead of truly independent assessments, which could play a useful role in budget decisions, the impact statements have been bland products of an inter-agency consultation process in which the Arms Control Agency, which prepares the first drafts of the statements, has had the weakest voice and the Department of Defence the strongest. The function envisaged by the Congress is a valuable one, but for it to be performed with some independence of mind, it should not be lodged in the Executive Branch, as it now is, but rather it should be shifted to the Office of Technology Assessment (*OTA*), an organization which is maintained and funded by the Congress and which has a more independent and often sceptical view toward weapons projects it is called on to finance than the Executive Branch which is advocating the research on these weapons.

Ycl it is posing high demands to expect a government to discipline itself in this matter more or less on its own. One answer may lie in consultative machinery which provides an authorized channel for cautionary comment from other governments.

11.5 Controlling Innovation in Nuclear Weapons

Two possibilities recommend themselves, one for nuclear weapons, and one for conventional weapons.

For nuclear weapons, it should be possible without the need for formal treaty agreement to establish a committee on new nuclear weapons technology with participation by all five declared nuclear weapons states, including, of course, the Soviet Union (or its successor states). Establishment of such a committee should follow the success of efforts by the United States and Soviet Union in cooperation to bring the other three declared nuclear powers—France, the United Kingdom, and China—into a regime of controls on the size and nature of their nuclear forces, as well as of efforts to make active use of the *UN* Security Council to bring about a more effective non-proliferation regime.

An institution to discuss technological innovation was foreseen at Soviet urging, in the unratified *SALT* II Treaty. In this case, it would be used to discuss the implications of new technology for the force balance between the five countries. This will be particularly desirable when, in due course, as seems probable and desirable, *US* and Soviet warhead levels art cut radically from their levels after ratification and implementation of the *START* Treaty. Especially at a low level of nuclear tor conventional) forces, important weapons innovation, by threatening to give the innovator real advantage, could have an extremely destabilizing effect. If the level of *US* and Soviet nuclear forces goes down drastically, if efforts arc made to make the *UN* Security Council a more effective vehicle for non-proliferation controls, and if all five nuclear powers accept a system of agreed low warhead levels (say, 1,000 for the *US* and *USSR* (or what has replaced it), and 700 for the other three slates), it will be necessary to avoid military technological innovations on the part of any one participant which could unbalance the relationship among them.[2] This is especially necessary if the five are to cooperate actively in Security Council peacekeeping and non-proliferation activities.

Here, some technological innovations might be relinquished for a different motive than in the past, not because of their possible effects in worsening relationships with an adversary and in intensifying military competition, but because of their potential for inhibiting cooperation among collaborating powers. It is also more possible in this environment than it was in the first half of the 1980s to imagine actual application of President Reagan's then rather implausible suggestion (in connection with *SDI*) that the United States could share some new defence technology with the Soviet Union successor States and other nuclear powers.

---

[2] An example is the new technology on laser-isotope separation, described in another chapter of this book by Annette Schaper, Wolfgang Liebert, Wim A. Smit, and Boelie Elzen: "New technological developments and the non-proliferation regime. Re-directing and constraining R&D: the case of laser fusion, laser isotope separation and the use of highly enriched uranium", in: Hans Günter Brauch, H. van der Graaf, John Grin and Wim Smit (Eds.): *Controlling the Development and Spread of Military Technology* (Amsterdam: VU University Press, 1992): 121–138.

11.6 Conventional Forces

New measures to deal with innovation in conventional forces might be based on two Confidence- and Security-Building Measures approved at Vienna in November 1990. These are the measure on Information on Plans for the Deployment of Major Weapon and Equipment Systems (12) and the measure on Information on Military Budgets (14). It could be argued in this context that the combined effects of political developments (including negotiated Soviet force withdrawals), an expanded *CSBM* regime, and the *CFE* Treaty (assuming ratification) have created a new military balance in Europe that should not be disturbed by major innovations in military technology. The *CSBM* calling for pre- notification of plans for deployment of major weapons systems was intended by its German originators as a measure to ward off the destabilizing effects of sudden deployment of major new weapons systems by pre-notifying that deployment.

But the measure was cut back in scope during the alliance consultation process and East–West negotiations. It now refers only to the types of weapons reduced or limited by agreement in the *CFE* Treaty—tanks, helicopters, armoured combat vehicles, heavy antitank missile launchers, artillery pieces, and armoured vehicle launched bridges—rather than to innovation in other weapons systems, including those utilizing new principles, like rail guns, and it requires only advance warning of a year. This measure should be developed further the warning period should be increased to 2 years. The scope of the measure should be broadened to equipment which represents a major innovation in some respect—firepower, armour, range, or mobility—or which uses physical principles other than those common in already deployed weapons.

It may be possible to push the obligation to exchange information on this subject to a point still earlier in the process of development of weapons systems, for example to the field-trial stage, and to arrange some observation of the trials by *CSCE* member governments. (It should be borne in mind that this measure covers only the *CSCE* area for security measures, the area from the Atlantic to the Urals. To become more effective, its coverage would have to be extended to the rest of the—former—*USSR* and to the home territory of Canada and the United States.)

The second existing *CSBM,* that providing for exchange of defence budgets, should be developed so that it would provide a basis for identifying and discussing innovative weapons systems early in their cycle of development. To achieve this, agreement should be reached on a new *CSBM* which would provide for periodic, separate discussion of military budgets, independent of the annual review of *CSBMs* to be carried out at the Conflict Prevention Centre, starting in November 1991, on the basis of exchanging data on defence budgets considerably more detailed than the brief summary, prepared according to *UN* format, provided for in the present measure. Even so, this would not prevent the continued existence of some secret or 'black' projects. New measures along these lines might be adopted in connection with the next Helsinki follow-up conference beginning in March 1992.

In the course of time, it might be possible for individual governments in reporting innovations to include their own impact statements, like those prepared by the United States Arms Control Agency. In time, it might be possible for such statements to be prepared by *CSCE* institutions like the Conflict Prevention Centre.

The fact that much research on weapons innovation is conducted in secret -an example is Stealth technology—will limit the effectiveness of multilateral discussion of weapons innovation. But even such research is often more broadly revealed in later stages, when it could be discussed. The habit of multilateral discussion could in time reduce secrecy and increase transparency in this field.

If the procedures recommended here were adopted, it could make it possible to discuss major weapons innovations several years before their deployment. This discussion would have several potential benefits: it would strip new deployments of the mystique and shock effect which during the Cold War elicited so much worst-case analysis, fuelling competitive weapons development. By providing an opportunity for criticism by other governments, and also for statements as to the potential counter-measures which deployment of the new system might elicit, perhaps reflected at some stage in public media. These procedures might slow the pace of technological innovation -not a dramatic achievement, but an important one. If a proposed innovation is sufficiently important, with the potential for disturbing collaborative arrangements among the five nuclear powers or among the CSCE states, it may be possible to reach some agreement among participants that it will be delayed or even not pursued.

Jonathan Dean visiting the People's Congress in Beijing during a conference on "Security, Arms Control and Defence Restructuring in East Asia", co-organized by COPRI, UNIDIR and the Chinese People's Association for Peace and Disarmament in Beijing in December 1996. In the middle is Dr. Björn Möller, then deputy chairman of AFES-PRESS. *Source* Personal photo collection of the Dean family

Chapter 12
NATO Enlargement: Coping with Act II (1999)

Coping with Act II

The Founding Act on Mutual Relations, Cooperation and Security between NATO and the Russian Federation, the document of understanding signed by NATO and Russia in Paris in May 1997, is a good idea on its own merits.[1] Even though conflicting U.S. and Russian interpretations of its content undermine the act's value, it is nonetheless worthwhile to establish a systematic means of communication between NATO and Russia and to attach Russian military officers to many NATO staffs.[2] But NATO enlargement—whose negative impact the Founding Act was intended to cushion—remains a bad idea.

If NATO enlargement were to stop with the admission of the three Central European candidate countries—Poland, the Czech Republic, and Hungary—that were invited to apply for membership in July at Madrid, the Founding Act might be helpful in persuading Russia to come to terms with that development. Then, whether NATO enlargement was a good idea or a bad one, we could all get back to business.

12.1 Prospects for Further Rounds of Expansion

But the question of NATO enlargement will not end with the admission of three new members. The Clinton administration has made clear that it plans to push for further expansion. Two days after the signing of the Founding Act, Secretary of State Madeleine Albright said at the NATO ministerial meeting in Sintra, Portugal, "We must pledge that the first new members will not be the last and that no European democracy will be excluded because of where it sits on the map That is why it is essential that NATO begin a new phase of dialogues with the aspiring

[1] This text was first published as "NATO Enlargement: Coping with Act II", in: Ted Galen Carpenter, Barbara Conry (Eds.): *NATO Enlargement—Illusions and Reality* (Washington, DC: CATO Institute, 1998): 121–127. The permission to republish this text was granted on behalf of the CATO Institute by Ms. Eleanor R. O'Connor on 18 March 2014.

[2] Madeleine K. Albright, "Statement at North Atlantic Council Ministerial Meeting, Sintra, Portugal, May 29, 1997," Office of the Spokesman, U.S. Department of State.

H. G. Brauch and T. Grimwood (eds.), *Jonathan Dean*,
SpringerBriefs on Pioneers in Science and Practice 19,
DOI: 10.1007/978-3-319-06662-2_12, © The Author(s) 2014

countries after Madrid."[3] NATO had already pledged itself in December 1996 to remain open to die accession of additional members once the first group of candidates had been accepted. At the Madrid summit alliance leaders actually specified five more candidates—Romania, Slovenia, and the three Baltic states—that will probably be invited to negotiate accession agreements in April 1999, when the Madrid candidates will be admitted to full NATO membership.[4]

Today, there are nine candidates for NATO membership in addition to the three Madrid entrants: Albania, Bulgaria, Estonia, Macedonia, Latvia, Lithuania, Romania, Slovakia, and Slovenia. Austria, Finland, and Sweden are also considering applying. The day after the Madrid summit meeting, the ambassadors of the three Baltic republics—Estonia, Latvia, and Lithuania—were on NATO's doorstep, clamoring for admission in the next round. The prospect of NATO membership for those three small states, which would bring the alliance directly to the borders of Russia, will confirm Russia's worst fears and renew active controversy over the entire enlargement issue.

The three Baltic governments are unlikely to stop pressing their candidacy—visibly and noisily—until they are admitted. The three governments have already indignantly rejected a 'charter' like the one developed for Russia; they want only full NATO membership, and they will not stop agitating for it until they get it. Senate Republicans have introduced a bill endorsing the candidacy of the Baltic states, and Majority Leader Trent Lott has pledged his support.[5] The risk that the Republicans could accuse the Clinton administration of yielding to Russian pressure and dragging its feet over this question in the 1998 and 2000 election campaigns will ensure visible and energetic administration support for the candidacy of those three countries. Baltic candidacy is also being pushed by the Scandinavian countries—even Finland and Sweden, who are not now members of NATO—in the interests of Baltic solidarity and of bolstering their own security situations.

The motives of the three Baltic governments are wholly understandable: they are seeking at any cost to ensure their own security in the face of Russian political opinion that barely concedes their independent existence. Unfortunately, their geographic situation means that NATO could not defend them militarily against Russian pressures except by threatening to use nuclear weapons. Given that the issue of NATO membership for the Baltic republics will certainly elicit a hostile reaction from Russia, the situation that will be created is not one that a farsighted American government should get itself into unless the stakes are the life or death of the United States. In this case, those are not the stakes. Instead, they are the historically understandable desire of three small countries for reassurance, which can be provided in ways other than this potentially suicidal one.

---

[3] Statement of the Secretary General, NATO Press Office, July 8, 1997.

[4] Trent Lott, "The Senate's Role in NATO Enlargement", Washington Post, March 21, 1997.

[5] Lionel Barber, "No Turning Back from Brave New Europe," Financial Times, July 17, 1997.

Ukraine will be satisfied—for a time—with the charter it received at the Madrid summit. However, if the Baltic states get into NATO or seem to be succeeding in their quest, Ukraine will likely become a candidate, too. None of the other candidates for NATO membership is a former Soviet republic or adjoins Russian territory. Thus, the candidacies of the other states will not cause so much reaction from the Russian political elite—although their candidacies should cause Americans to ask many questions about whether the United States should extend a security guarantee to so many countries.

All of this means that the negative aspects of NATO enlargement will not end with the accession to NATO of Poland, Hungary, and the Czech Republic. Instead, the NATO enlargement issue will cause continuing friction with Russia. It has already held up Duma ratification of START II, and it will seriously impede Russian cooperation on other valuable arms control projects. East European states will be caught up in a political maelstrom of rival candidacies. The needless and destabilizing remilitarization of Eastern Europe will continue, at a cost that will greatly exceed the $30 billion the Clinton administration has estimated for the Madrid candidates. With at least five more probable members, those costs will at least double. And if all 16 countries named here should become members of NATO, the costs would be from three to five times the administration's estimate.

In the minds of its leading proponents, the NATO enlargement project is designed to prevent a repetition of the situation that led to two world wars—the existence of a belt of small, weak East European states caught between competing major powers. But the enlargement project may well create the very situation it is intended to prevent: quarrels between two power blocs over weak buffer states. In particular, the hostile reaction of the Russian political elite on this issue will long mold the views of the average Russian on foreign affairs. The NATO enlargement controversy will revive paranoid theories of hostile encirclement, which have played such negative role in Russian history, and dispose Russians to a dangerous revisionist attitude toward the post-Cold War settlement in Central and Eastern Europe. The situation will be handcrafted for an unscrupulous political demagogue.

12.2 Alternatives to NATO Enlargement

There are better alternatives to NATO enlargement. They include strengthening the Organization for Security and Cooperation in Europe, already a pan-European organization, to become more effective. They also include building the Partnership for Peace into a more serious program. Above all, they include pressing the European Union—to which Russia does not have the same built-in, automatic hostility as it has to NATO—to move more rapidly toward its own enlargement in Eastern Europe. Negotiations on that subject are scheduled to begin in 1998. The European Commission has already specified the Czech Republic, Poland, and Hungary, as well as Slovenia and Estonia, as the first group of enlargement candidates, and if all goes well, the first three countries will be admitted to the EU by 2003 or so, only a few years after their possible admission to NATO.[6]

From the outset, the real interest of the East European countries has been in membership in the EU and in the long-term economic support such membership promises. For many East European states, NATO membership is only an interim surrogate for the real goal. Opinion polls in the main candidate countries show higher public interest in EU membership than in membership in NATO.[7] Negotiations for membership in the EU are slow paced and may take several years. However, there is no crisis in Eastern Europe that requires haste or rapid solution. If membership in the EU for some East European states takes several years to work out, then so be it. One of the most questionable justifications for the NATO enlargement project is the argument that a pending crisis in Eastern Europe makes immediate expansion of a military alliance—NATO—a necessity. There is plenty of time to develop a more productive answer to East European concerns. But in the present rush to enlarge NATO, Western legislators and populations—who will have to assume the costs and risks of enlargement—are being stampeded into unwise decisions.

Logically, the serious problems associated with NATO enlargement should occasion serious debate over the entire project when the amendment of the North Atlantic Treaty to include Poland, Hungary, and the Czech Republic is presented for ratification to the U.S. Senate and other NATO legislatures. It is to be hoped that the Senate will conclude that the entire project should be kept on hold while other alternatives are more thoroughly studied. But, whatever the Senate's decision on the candidacy of Poland, Hungary, and the Czech Republic, part of its decision should be a firm requirement for a moratorium on further NATO enlargement.

The administration should use that moratorium to consult with other NATO governments to develop a special program for the Baltic states that takes into account their sensitivities, those of the Russians, and the Baltic states' exposed geographic position and offers them reassurance outside the NATO framework. Specifically, the administration should work with the EU to develop for the Baltic states a special, intensified form of EU association with most of the benefits of EU membership—an approach that can go into effect rapidly. It would be in the political interest of all EU countries to avoid the acute friction with Russia that would result from the active NATO candidacy of the Baltic republics by making available to them a more constructive alternative, such as close EU association.

---

[6] George Cunningham, "EU and NATO Enlargement; How Public Opinion Is Shaping Up in Some Candidate Countries," NATO Review, May–June, 1997.

[7] Vaclav Havel, "NATO's Quality of Life", New York Tunes, May 13, 1997; and Jim Hoagland, "At Center Stage for Havel: NATO," Washington Post, May 9, 1996.

12.3 Toward Comprehensive Enlargement

If, in fact, despite the risks, the legislatures of all present NATO members do approve the membership of the Madrid candidates, the extended moratorium on further candidacies does not prove feasible, and the administration and the EU fail to develop a special program for the Baltic states, only one effective possibility of limiting the damage from NATO enlargement will remain open. The basic problem of NATO enlargement is that by including some countries and excluding others, especially Russia, it establishes a new dividing line in Europe. What is clearly missing is a comprehensive all-European solution. The Germans have an expression that would apply in such circumstances: *Wenn schon, denn schon*—"If you are going to do something that may be questioned, then do it right and go all the way."

In these circumstances, the Clinton administration should draw the logical conclusion from its project to enlarge NATO. Acceptance of the first group of candidates, if it comes, will mean that the United States and other NATO members are committed to NATO as the prime security organization of Europe. No serious alternative—including the OSCE and the EU with its military arm, the Western European Union—will then remain for that role. Both the logic and the politics of the situation will then require that the United States go the next step and open NATO membership to all democratic European states.

Specifically, the administration should prepare a 20-year plan for the comprehensive enlargement of NATO, a plan that would provide for NATO membership for all of the candidates, including the Baltic states (and Ukraine if it so desires), and also, at the end of the 20-year period, of Russia itself. The plan should be detailed and credible.

Russia has shown interest in membership in NATO on several occasions. Soviet president Mikhail Gorbachev suggested it; Russian president Boris Yeltsin suggested it. Early this year Prime Minister Viktor Chernomyrdin suggested at the Davos economic conference that Russia become a member of the NATO council. Perhaps Russian membership could be ultimately achieved by amalgamating the new Founding Act's permanent joint council with NATO. Regardless of the method, if Russia can be genuinely convinced there is a solid plan and a solid prospect for its own ultimate NATO membership, it is less likely to object in the interim to membership even for the Baltic states and Ukraine, much less the other candidates.

There are some problems with such a solution, although none of them is as large as those caused by the NATO enlargement project itself. At present, one function of NATO is to ensure against a resurgent Russia—a not unreasonable mission, given Russia's unstable polity. That is why this proposal for comprehensive enlargement suggests a 20-year period for Russia to settle down. Russia will either do so in that time period or will at some point during that period take such a negative course as to make its entry into NATO impractical.

Russia is also characterized by some Europeans, for example, by President Vaclav Havel of the Czech Republic, as a vast Eurasian country—too large and too alien to be absorbed into NATO without changing NATO's own nature.[8] But Russia is a European country, and if it becomes a functioning democracy, its other characteristics will not be so important. True, the addition of 10 more members, including Russia, would probably change the nature of NATO. However, over time, as Havel himself has suggested, NATO should change from a military alliance directed against a specific enemy to a collective security organization with the capability of mounting military missions supported by coalitions of willing members 7.

There is another practical concern. The primary reason many members of the Senate today favor NATO enlargement is suspicion of Russia—Russians are right when they conclude that many of those legislators intend the enlargement of NATO as an anti-Russian measure. If a plan to make NATO membership genuinely comprehensive were publicly advanced now, such legislators would probably reject it. They might even vote against the first tranche of NATO enlargement if it were seen as part of a comprehensive plan of enlargement that would include Russia at a later date.

But we are now only at the beginning of the wide public debate the NATO enlargement project calls for; the potential long-term costs of the project are becoming more evident, and the need for remedial action will also become more clear. Moreover, it is the administration's obligation, having conjured up the NATO enlargement project, to cope with those factors and to reach a decision on the timing of a proposal for comprehensive enlargement of NATO.

The task of devising a new security architecture for Europe that the administration has spoken of so frequently will not be completed until there is an enduring, effective place in that architecture for all European states, including Russia. That is what is missing from the NATO enlargement project in its present divisive and exclusionary form. If the administration fails to carry out this task properly, we will all pay the costs.

[8] Ibid.

Chapter 13
Ten Years After the Wall: Trends in Post-Cold War U.S. Security Policy (1999)

Ten years have passed since the inaugurating act of the post-Cold War era–the demolition of the Berlin Wall.[1] The "10 Years After the Wall" symposium held at the American Academy of Arts and Sciences in Cambridge in November of 1999 examined the questions of how and under what influences U.S. military and security policy has changed during the past decade. More than a simple retrospective, however, speakers looked to the future, assessing the direction of current policy, and then reflected on "the paths not taken."

For each of the topic areas, the questions animating the proceedings were: Where has security policy been in the recent past, where is it going, and why? How well is the U.S. adapting to the challenges and opportunities of the new era? The symposium sought to engage both past choices and present options, paying attention to how the security policy debate has reflected differences over policy goals and priorities, different assumptions about the security environment, and different evaluations of the policy instruments and resources available.

One hundred and forty people attended the symposium, with a majority of the participants students or teachers of security studies and international affairs, leaders of advocacy groups, and other specialists, including military officers and members of the press.

13.1 Jonathan Dean, Speech as Given on 11 November 1999

This afternoon, I will describe post-cold war U.S. security policy in three major areas, evaluate how well the U.S. has carried out these policies, and suggest future prospects. The areas are: U.S. security cooperation in Europe; U.S. policy regarding the UN; and U.S. policy on nuclear and conventional arms control.

As we conduct our evaluation today, we should keep firmly in mind the remarkable non-violent ending of the cold war: The end of the NATO-Warsaw Pact confrontation, the reunification of Germany, withdrawal of Soviet forces from

[1] A 2 Day Symposium was held on November 11–12, *1999* at the American Academy of Arts and Sciences, Sponsored by: *Project On Defense Alternatives,* Commonwealth Institute Cambridge, Massachusetts. This text was not published previously.

H. G. Brauch and T. Grimwood (eds.), *Jonathan Dean*,
SpringerBriefs on Pioneers in Science and Practice 19,
DOI: 10.1007/978-3-319-06662-2_13,

Europe, and peaceful dissolution of the Soviet Union itself. It took great skill to avoid the bloodshed that these developments could easily have brought. These were enormous achievements, for which we must give credit to the Bush and Clinton administrations. The nuclear weapons of the Soviet Union were also gathered into Russia and there has been an underfunded but well-conceived program to cope with the problems of loose nukes and migration of Russian WMD experts to foreign employ.

There have been some other important gains for U.S. security policy in this period. Saddam Hussein's aggression against Kuwait and his violation of the Non Proliferation Treaty was frustrated. North Korea was blocked from moving to substantial nuclear weapons capability. There were partially successful UN actions in Cambodia, Haiti, Bosnia, Somalia, Kosovo and East Timor. The Non-Proliferation Treaty was extended indefinitely. The Chemical Weapons Convention was concluded and entered into force. A Comprehensive Test Ban Treaty was negotiated and signed. NATO was saved from oblivion and decline by enlargement and by involvement in Bosnia and Kosovo.

Some may not consider NATO's survival a positive achievement, but, after a 50 year fight against political evil, it was not at all unreasonable to expect that developments like these would open a new and better era, a new world order, as President Bush put it at the UN after the Gulf War ended.

Yet this did not happen. For the United States, the main factor preventing this new era was not so much poor leadership or poor policy—the main topics of our discussion here—but something quite different and outside the framework of good or bad policy. It was the unexpected revival of traditional American isolationism, transmuted into the new phenomenon of unilateralism. Today, the current dominance of the unilateralist trend in the continuing struggle between groups in American society supporting concepts of strong reliance on multinational cooperation and multilateralism, and the groups supporting the idea that the U.S. should rely mainly on its own strength threatens to cost the United States much of the prestige and influence that it gained from the peaceful end of the cold war. It threatens to bring deterioration in U.S. relations with Russia, China, and Europe, to cause failure in using the UN as an effective instrument for furthering U.S. values, and, perhaps worst of all, to cause the collapse of nuclear disarmament. A lot has been said on this topic in past weeks, but it deserves our continuing close attention. In fact, it should have received the unwavering attention of the Clinton administration for the past 5 years.

13.2 Europe

To turn to our three security policy topics: In Europe, post-cold war American security policy has had three main objectives: To further the integration of Western Europe; to integrate Eastern Europe into the Western structure; and to integrate Russia into the Western circle. The first task has largely succeeded with the beginning of European Monetary Union. However, Germany's still incomplete unification and the move of the capital to Berlin has disjointed the Franco-German partnership that has driven European unification and has made it even more clear that the European Union is moving towards confederation, not federation.

The second objective, integration of the former Warsaw Pact countries of Eastern Europe into the West has partially succeeded. The difficulties of German unification and the resultant delays in EU enlargement to the East brought the use of a substitute mechanism, NATO enlargement, which frustrated the third objective, integration of Russia.

There are at least 15 more candidates for NATO membership, a fact which could keep the enlargement issue alive as a source of serious friction with Russia for decades to come. The integration process has also left out part of Yugoslavia, which Marshall Tito had taken out of the alliance with the Soviet Union. Yugoslavia has remained under fairly continuous authoritarian rule from the dawn of recorded history, with tribal and clan cohesion as the only dependable social foundation. Now, belatedly, these poor Balkan cousins of the European Union will have to be gathered into the family. Even Albania has gained political entree through its asylum for the Kosovars. Quite possibly, a 40–50 year process is involved here, but European television and other media will probably keep the issue alive during that period and the process will continue fitfully.

With Russia, the first wave of attempted Western integration failed. Western governments put up insufficient money. There was no Marshall Plan organization of donor governments with the authority that was needed to overrule sovereignty-based defenses for theft, corruption, and the absence of accountability in Russia. The procedure the West used for integrating Eastern Europe, the enlargement of NATO, alienated Russia. American military power used in Iraq and in Kosovo over Russian objections, and U.S. plans for nationwide missile defense of the United States brought fear and more alienation.

It would have been extremely difficult under any circumstances to conduct a Western policy that could avoid intensifying Russia's pain over its loss of international status. Efforts were made, but in the cases I cite, Russian sensitivities were trampled on. Now, the job of integrating Russia will have to be done the slow way over a 50 year period through German-led European Union investment and possible ultimate integration.

The missed opportunity means that Russia may go through serious detours—fascism or military rule, or the two combined. The Caucasus area has already become Russia's Balkans. The worst victims of this process will continue to be the Russian people, but there will also be increased danger of military clashes with the outside world. Making NATO membership universal for all potential candidates, including Russia, would ease the continuing friction of that process, but the harm seems to have been done in terms of creating enduring Russian hostility toward the United States, although fortunately less for the European Union states.

As regards the long-term future of Europe's security organizations, ultimately, I see the European Union slowly absorbing both the OSCE, NATO and WEU.

13.3 The United Nations

United States post-cold war policy toward the United Nations has also been a failure, at least for the short term, in the sense of failure to exploit the huge opportunity that arose at the end of the cold war for building this organization into an effective mechanism for peacemaking. America's enormous post-cold war prestige and political, economic and military power could have provided great leverage for this effort.

In the Gulf War, Haiti, Bosnia and Kosovo, the United States was able to use its military capability, joined with that of its allies, with considerable military effectiveness. But this has been done at the cost of a serious drop in the standing of the UN in United States political opinion, and to some extent, Western political opinion, a decline which came as the U.S. blamed the UN for its own errors in Somalia, Rwanda, and Bosnia. This development has also brought a drop in U.S. prestige.

Humanitarian intervention was justified in Kosovo. The military means used appear to have cancelled out the humanitarian gains. But other uses of military force, like use of combined ground and air forces, if they had been politically feasible, would probably have had similar effects. Possibly, a proposal for a UN-led peacekeeping force might have been more acceptable to Milosevic, and might have cut the duration of the NATO bombing. But by this time, American political opinion, including administration opinion, had so thoroughly trashed the reputation of the United Nations that it was no longer politically feasible for the administration to propose the use of UN peacekeepers in Kosovo.

Part of the problem here is the failure of the U.S. administration to present humanitarian intervention as a U.S. national security requirement, justified when governments violate their stewardship for the welfare of their population with serious human rights abuses or, put another way, when governments abuse the sovereign powers entrusted to them by their populations. However, instead of presenting the prevention of human rights abuses as a recognized and legitimate national security interest of this country, the administration has presented it as an optional extra to much higher ranked national security interests. As a result, the administration has been obliged to laboriously try to explain in the individual case why the U.S. is getting involved after it has repeatedly said that these humanitarian crises were not important. It should be the other way around: When appropriate, the administration should explain why action by others or by the UN itself will make U.S. action unnecessary to maintain a clearly established policy of humanitarian intervention.

Nonetheless, controversy over Kosovo, East Timor and the many difficult cases like Liberia, Sierra Leone, Angola and Sudan has focused attention on the right place, on conflict prevention. Numerous actions can be taken in this field at the UN and elsewhere to improve capability for conflict prevention.

As many of you know, Randall Forsberg, Saul Mendlovitz, and I have proposed the Global Action to Prevent War project. Among other things, the Global Action program urges establishment of a professional mediation corps at the UN, and of a conflict prevention committee in the General Assembly. It urges the possibility of an understanding among permanent members of the Security Council to move into a proactive conflict prevention role and to limit the use of the veto for that purpose. We propose establishing a standing UN police force of 2–5,000 men and women which could be rapidly deployed to many areas where efforts to gain entry for military peacekeeping forces would cause problems. There are many other specific proposals in the Global Action program. They are on our website at www.globalactionpw.org.

13.3.1 Disarmament

Turning to the issue of disarmament and arms control, following rejection by the United States Senate of the Comprehensive Test Ban Treaty, there is wide agreement that the present situation of nuclear arms control and disarmament is abysmal.

I believe we are at the end of a 40 year era of Russian-American nuclear arms control. Both countries have shifted their positions significantly away from cooperation on nuclear reductions toward national self-reliance.

Instead of a situation where we could look forward to a relatively continuous process of incremental steps of nuclear arms control, we appear to be on the verge of deployment of nationwide missile defense, collapse of the ABM Treaty, and even possible collapse of the NPT regime.

Rather than trying to paste together the old policy, I believe it is time to move to an approach that can cope with these dangers, bring Russia back into serious negotiation, and make China a full player, as well as establishing a role for India, Pakistan and Israel, and in this way compensate to some degree for declining American motivation.

To carry out this approach, the United States and Russia would agree as a first step to reduce to 1,000 warheads, real count.

All five weapon states would then immediately begin to negotiate comprehensive information exchange on their holdings of warheads and weapons-grade fissile material, reduction of all nuclear warheads in their possession to no more than 200 warheads each, and transfer of fissile material for weapons to monitored storage following verified agreement to end its production. Remaining permitted warheads would be separated from delivery systems. Both would be stored under international monitoring on owner state territory in multiple, dispersed sites that could have site defense against air or missile attack.

The three de facto nuclear weapon states—India, Pakistan and Israel—would be urged to store their nuclear weapons materials and delivery systems under similar conditions. No realistic proposal for deep cuts in nuclear arsenals can omit coverage of the de facto weapon states.

Details of the program I have just sketched out are in a newly published book, *The Nuclear Turning Point,* edited by Harold Feiveson and published by the Brookings Institution.

If it were actually implemented, the program I just described here would cover all known weapon states and would eliminate the worst dangers from current nuclear arsenals, including pre-emptive attack, accidental or unauthorized launch, and threats of use of nuclear weapons. It would sharply cut risks of diversion of nuclear weapons and fissile material and of nuclear proliferation. Moreover, it would set the stage for the first serious discussion with the weapon states of complete elimination of nuclear weapons. This approach might ultimately be attractive to Russia and to China because it would deal with their concerns about American nuclear superiority and missile defense, and also attractive to India and Pakistan, because it would give them an active role in disarmament politics.

I am hoping for favorable statements about this approach from non-weapon states in the NPT review conference—from the New Agenda Coalition, from the non-aligned movement, and especially from China, Russia, and also from India and Pakistan outside the NPT. A broad coalition like this could bring the United States to seriously consider the coalition's proposals for nuclear disarmament.

We should also introduce a program of conventional disarmament, including the disarmament among the major powers that is required to make nuclear disarmament feasible. The Global Action program has a practical proposal for this.

In conclusion, I would point out that one common characteristic unites U.S. post-cold war policy on all three of the subjects I have touched on here: it is the continuing ambivalence and inconsistency of U.S. policy arising from the unresolved confrontation between the concepts of multilateralism and unilateralism.

In European security policy, we see this ambivalence in U.S. emphasis on NATO and U.S. reluctance to build OSCE and the Western European Union and the European defense identity within NATO. NATO enlargement itself was a reaction to mistaken administration fears of a tidal wave of American isolationism. Instead, the wave of unilateralism that actually took place approves NATO as the only good U.S. foreign commitment.

We see this ambivalence and incoherence in the refusal to follow the common sense course of building the UN, a tiny organization whose current budget and personnel total are smaller than those of my home county, Fairfax County, Virginia, into a more effective mechanism for preventing conflict. The administration has not even succeeded in paying the costs for peacekeeping operations that the U.S. itself approved in the Security Council. In the end, this struggle also controlled the nature of the U.S. response in Kosovo, with its reliance on NATO and on bombing. We see its effects in disarmament, where the Senate majority has rejected the Comprehensive Test Ban Treaty and voted in favor of a free hand for American military and technological capacity and of taking our chances in a proliferated world. And we see it in the pressure of the majority party in the Congress to push ahead with nationwide missile defense in the United States and with theater missile defense on the periphery of China, and the administration's surrender to that pressure.

Perhaps it should not have been one of the major surprises of the post-cold war period that traditional isolationism should suddenly erupt in the 1994 congressional elections in its new guise of unilateralism. After all, this strand of American thinking, historically based, and nourished by American exceptionalism, had been repressed as unpatriotic for at least half a century of World War II and cold war. It was perhaps inevitable that these dammed up sentiments would erupt as soon as the cold war cap was removed, and understandable that they would break through in changed form.

Fifty years of television coverage of the outside world has meant that it is no longer possible for any American to wholly ignore the existence of the 96 % of the human species outside our borders. Old-fashioned isolationism has passed on the torch. The lesson of overwhelming American military power has also been absorbed in this new phenomenon of American unilateralism. Unilateralism no longer rejects the outside world as such, but opposes a position of seeking to cooperate with it.

In actuality, unilateralism is classic nationalism. We cannot expect that this spurt of American nationalism will seep away on its own. After all, it has been an almost universal post-cold war phenomenon all over the world. So we cannot expect the complete elimination of these trends. But we should not accept their easy dominance over American security policy. We can rightfully expect that U.S. national leadership will effectively make clear to the American public the destructive effects of unilateralism on American security, and we can hope that each of us will vigorously engage in this debate, which is vitally important for the future of our country in coming decades and quite possibly for the future of the world.

Chapter 14
Future Security in Space: Treaty Issues (2002)

14.1 The Present Situation: Anarchic Coexistence of Military and Commercial Assets

The present rather anarchic coexistence of military and commercial assets on the space frontier may come to an end within the next 5–10 years.[1] This could happen if a new stage—weaponization of space—begins and is followed by the first space weaponizing power's promulgation of its own rules of the road for space. Or, it could happen if the present near anarchy is replaced by a new international rule of law, whether via a formal treaty or political agreements.

In any event, there cannot be much doubt that the weaponization of space will begin in the foreseeable future unless it is hindered by organized, effective international opposition or the current U.S. administration is replaced by an administration willing to cancel projects for weaponization of space in the face of considerable opposition from domestic proponents of weaponization.

This issue is not a theoretical one. Development of two specific space-based weapons, a kinetic kill weapon and a space-based laser, is official policy of the U.S. administration as part of its missile defense project. It is not a priority, but it has a specific development program and a budget of many millions projected over

[1] This text was first published as "Future Security in Space: Treaty Issues", in: Occasional Paper No. 10, James Clay Moltz (Ed.): Future Security in Space: Commercial, Military, and Arms Control Trade-Offs Special Joint Series on Missile/Space Issues (Monterey, Ca.: Center for Nonproliferation Studies—Southhampton, UK: Mountbatten Centre for International Studies): 3–7; the whole report is at: http://cns.miis.edu/opapers/op10/op10.pdf. The permission to republish this text was granted by Mr. Jon Wolfsthal, Deputy Director, CNS on 25 March 2014.

H. G. Brauch and T. Grimwood (eds.), *Jonathan Dean*,
SpringerBriefs on Pioneers in Science and Practice 19,
DOI: 10.1007/978-3-319-06662-2_14, © The Author(s) 2014

the next several years. The first space tests are scheduled to take place in 5–6 years, with deployment 5–6 years thereafter. This action is backed by an official doctrine calling for U.S. supremacy in space and doing what is needed to achieve that supremacy.

It is sometimes argued that space weapons won't work and therefore that concern over this subject is misplaced. This is a misleading argument. There are real doubts that some forms of missile defense will be effective, but this does not prevent justifiable concerns over the effects of U.S. deployment of nationwide missile defenses. The deployment of the first weapon in space within the next 5–6 years will cross the fateful line into the competitive weaponization of space.

It can be argued that, before these two missile defense weapons are orbited, the weaponization of space will begin with the deployment of the first ground-based interceptor missile in Alaska in 2004. It will be possible to use this system to destroy or incapacitate satellites moving in fixed orbit more easily than destroying an incoming missile. These interceptors will in practice be anti-satellite weapons, or ASATs. Beyond this, in May 2002, the House Armed Services Committee placed an unsolicited sum of $7 million in the fiscal 2003 budget for the Defense Department calling for continued development of a ground-based ASAT.

As weaponization appears to loom closer, there have appeared several suggestions for partial measures: "rules of the road" or confidence-building measures that would restrict weaponization or mitigate its effects if it comes. Some examples include: keep-out zones around satellites; non-interference and protection for manned space flight and manned space vehicles; and pre-launch notification and verification. One recent article suggests a mixed regime of some weaponization and some reassurance measures.[2]

Regardless of their motives, those advocating mixed regimes appear to assume that the weaponization of space is inevitable and that we might as well make the best of it. Must we take this outcome for granted? It would be far more desirable to reach agreement on a treaty prohibiting the weaponization of space, while also assuring use of space for both military and civilian observation and communications, with their own rules of the road governing this non-weaponized situation. I do not think this possibility can or should be ruled out. Furthermore, it should be possible to enter into dialogue with the United States to negotiate widely beneficial confidence-building measures while leaving open for later discussion the question of whether space-based weapons shall be permitted or banned.

[2] James Clay Moltz, "Breaking the Deadlock on Space Arms Control," *Arms Control Today* 32 (April 2002).

14.2 Existing Legal Structure

In regards to the existing legal structure relating to space weapons, the principal relevant treaty is the 1967 Outer Space Treaty (OST), which prohibits the orbiting or stationing in space of weapons of mass destruction, but not other weapons. I will return to the OST in a moment. Five other treaties address outer space. They include: the Limited Test Ban Treaty of 1963, which prohibits nuclear tests (and any other nuclear explosion) in the atmosphere or in outer space; the Astronauts Rescue Agreement of 1968; the Liability Convention of 1972; the Registration Convention of 1976; and the Moon Agreement of 1984.[3] These last four treaties elaborate aspects of the 1967 treaty.

In addition, there are five relevant General Assembly resolutions: the Declaration of Legal Principles States in the Exploration and Uses of Outer Space (1963); the Declaration on International Cooperation in the Exploration and Use of Outer Space for the Use and Benefit and in the Interest of All States (1996); and also resolutions on Direct Television Broadcasting,

Remote Sensing of the Earth from Outer Space, and the Use of Nuclear Power in Outer Space[4]. Since June 13, 2002, when U.S. withdrawal from the ABM Treaty became effective, there is no longer a treaty prohibition against testing or deploying weapons in space other than weapons of mass destruction. Lt. General Ronald Kadish of the Missile Defense Agency has already ordered ground-breaking at Fort Greeley for the missile defense installation there.

A sixth treaty is relevant to space weapons. The concept of non-interference with national technical means of verification first appeared in the SALT I Treaty of 1972 and was taken over into the START I Treaty, which has been prolonged to 2009. Similar protections are imbedded in the INF Treaty and the Conventional

[3] The Agreement on the Rescue of Astronauts, the Return of Astronauts and the Return of Objects Launched into Outer Space (the "Rescue Agreement"), opened for signature on April 22, 1968, entered into force on December 3, 1968, 87 ratifications; The Convention on International Liability for Damage Caused by Space Objects (the "Liability Convention"), opened for signature on March 29, 1972, entered into force on September 1, 1972, 81 ratifications; The Convention on Registration of Objects Launched into Outer Space (the "Registration Convention") opened for signature on January 14, 1975, entered into force on September 15, 1976, 43 ratifications; The Agreement Governing the Activities of States on the Moon and Other Celestial Bodies (the "Moon Agreement"), opened for signature on December 18, 1979, entered into force on July 11, 1984, 9 ratifications (As of February 1, 2001)

[4] The Declaration of Legal Principles Governing the Activities of States in the Exploration and Uses of Outer Space (General Assembly resolution 1962 (XVIII) of December 13, 1963); The Principles Governing the Use by States of Artificial Earth Satellites for International Direct Television Broadcasting (resolution 37/92 of December 10, 1982); The Principles Relating to Remote Sensing of the Earth from Outer Space (resolution 41/65 of December 3, 1986); The Principles Relevant to the Use of Nuclear Power Sources in Outer Space (resolution 47/68 of December 14, 1992); The Declaration on International Cooperation in the Exploration and Use of Outer Space for the Benefit and in the Interest of All States, Taking into Particular Account the Needs of Developing Countries (resolution 51/122 of December 13, 1996).

Forces in Europe (CFE) Treaty. The intent of these measures is to preserve from attack or interference satellites involved in verification. As I read it, it would be a violation of the provisions on noninterference with national means of verification in the START I and INF treaties to use weapons against any early warning, imaging, or intelligence satellite and, by extension, against any ocean surveillance, signals, intelligence or communications satellite of the U.S. or Russia. This obligation was made multilateral in the CFE Treaty.

14.3 The 1967 Outer Space Treaty

The 1967 OST is based on "the common interest of all mankind in the … use of space for peaceful purposes." The treaty forbids the orbiting or stationing in space of weapons of mass destruction and prohibits the use of the moon or other celestial bodies for other than peaceful purposes. The treaty contains four explicit references to the peaceful use of outer space.

This language points to the fact that, during the 30-year existence of the OST, a powerful norm has emerged against the weaponization of space, for keeping armed conflict out of space, and for ensuring the peaceful use of space. This conclusion is documented by UN General Assembly resolutions each year for the past 21 years calling for maintaining peaceful uses of space and opposing its weaponization. Most of these resolutions have been unanimous and without opposition, although the United States and a few other governments have abstained. In the most recent version of December 2001, the General Assembly once again passed, by 156 votes for to zero opposed, a resolution calling for negotiation in the Geneva Conference on Disarmament of a treaty to prevent an arms race in outer space. This time, there were four abstentions to the resolution. The now customary trio of the United States, Micronesia, and Israel was joined by a fourth country, Georgia. The resolution asks all treaty parties to refrain from actions contrary to the peaceful use of outer space and calls for negotiation in the Conference on Disarmament on multilateral agreements to prevent an arms race in outer space.

These repeated, nearly unanimous resolutions, against which even the United States does not vote, are not only evidence for the existence of a norm against the weaponization of space. They also indicate a very widespread desire to expand existing multilateral agreements to make explicit a prohibition against all weapons in space.

Article IV of the OST prohibits placing in orbit around the earth any objects carrying nuclear weapons or other weapons of mass destruction. It also prohibits the testing and, I would argue, the deployment of any kind of weapon on the moon or other celestial bodies. There is no provision for verification. As is well known, the 1967 Treaty does not prohibit the orbiting in space of weapons other than nuclear weapons or other weapons of mass destruction.[5]

However, the OST is not without useful features relevant to the possible weaponization of space. Article VII makes treaty parties that launch objects into outer space liable for damage to the property of another treaty party—this is also spelled out in the Liability Convention of 1972. The Liability Convention foresees the establishment of a Claims Commission to determine the extent of liability for damage by the space objects of one country to the space objects or property of another state. Article IX of the OST provides for consultations if any treaty party believes an activity planned by another treaty party would cause "potentially harmful interference with activities in the peaceful exploration and use of outer space." Beyond this, the General Assembly could by majority vote request an Advisory Opinion from the International Court of Justice if either the peaceful uses language of the 1967 treaty or these two articles on liability and consultation come under dispute as the space-based component of the missile defense system advances.

In fact, requests for consultation or under Article IX, or also a General Assembly request for an advisory opinion, can and should come now to make world opinion aware of this issue before the damage has been done, and to motivate the United States government to study the issue seriously, including the possibility of rules of the road. The request for consultation under Article IX can come from any party or group of parties to the 1967 treaty. In addition, George Bunn and John Rhinelander point out in a letter to the editor in the June 2002 issue of *Arms Control Today*, that parties to the treaty could convene and issue an interpretation that the U.S. testing or orbiting of space weapons was contrary to the peaceful uses language of the treaty, in effect amending it to preclude any weaponization. The General Assembly could then pass a resolution endorsing this interpretation.

Presumably, Russia, the United States, or any state party to the CFE Treaty could also take legal action based on treaty provisions prohibiting interference with national technical means of verification. In addition, legal action could be taken in U.S. courts by U.S. commercial users of space satellites if these satellites were endangered by U.S. space weapons. In theory, other legal actions might be taken by private corporations at the Hague Court of Arbitration if the defendant state is willing to permit this. In short, existing space law provides numerous opportunities to remind the United States that weaponization of space could be a complex and difficult process, to make it worthwhile for the U.S. government to negotiate on confidence-building measures, or, if necessary, to block early weaponizing measures.

---

[5] Article IV of the 1967 Treaty states: "States Parties to the Treaty undertake not to place in orbit around the Earth any objects carrying nuclear weapons or any other kinds of weapons of mass destruction, install such weapons on celestial bodies, or station such weapons in outer space in any other manner." This language would appear to preclude orbiting weapons around the moon, but Article III of the 1979 Moon Treaty makes this prohibition explicit.

It is relevant to this subject that there have been press reports that the U.S. Defense Science Board has expressed interest in reexamining the possibility of using nuclear warheads for missile defense purposes. Explosion of nuclear weapons in the atmosphere or in space is explicitly forbidden by the 1963 Limited Test Ban Treaty (LTBT). Action by the Defense Department to carry out these plans would be a violation of the LTBT and would in addition lead to a multitude of suits and injunctions under the Liability Convention. The same applies to tests of the GALOSH missile defense system around Moscow, which continues to be armed with nuclear warheads.

14.4 Prospects for Expansion of the Legal Regime

Driven by U.S. moves to weaponize space, it appears likely that the legal regime covering space weapons will expand in the next several years for a variety of reasons.

One possibility, of course, is action to fill in the gap in the 1967 OST and to prohibit any weaponization of space. Russia and China have proposed comprehensive treaties prohibiting weapons in space. In 2001, China presented component concepts of a draft treaty on this topic in the Conference on Disarmament (CD) and in May 2002, China and Russia presented a joint draft in the CD. The prospects for a comprehensive treaty depend in large part on the willingness of China, Russia, and other large countries to undertake a major international effort supported by non-governmental organizations (NGOs) in the face of U.S. opposition. For this effort to succeed, China, Russia, and other major states, including European Union members, would have to conclude that preventing the weaponization of space is a top priority national interest, justifying the sacrifice of other important national interests in a stand-off with the United States. They do not appear to have reached this conclusion yet. However, they still may reach this conclusion at some future point, especially if the United States continues to make major increases in funding for its missile defense program. But, otherwise, it is plausible that these countries may seek some compromise with the United States that could result in a mixed regime of some weaponization and some regulation.

Second, the same difficulty of prevailing upon the world's most powerful state, the United States, to renounce categorically its plans to place weapons in space has already led to suggestions from think tanks and NGOs that concede some weaponization and argue for a mixed regime of partial weaponization backed by limits and controls. Think tanks and NGOs may be joined in this approach by commercial users of space that have thus far held back from taking a position in this controversy.

Finally, if the U.S. policy of placing weapons in space proceeds and results in de facto space supremacy for the United States, Washington itself is likely to attempt to reach agreement with other spacefaring governments on rules of the road, like keep-out zones, to blunt some of their criticisms, and, in the final analysis, to prevent situations where the United States may be forced against its will to use its space weapons.

The combination of all these pressures seems quite likely to bring additional agreements, both formal and informal, related to weapons in space and anti-satellite systems. The present quasi-anarchy of space law is likely to give way to a network of rules while preparations for weaponization proceed. This process and the discussion that will accompany it will also help to inform the U.S. electorate and perhaps incline it to back a preventive or corrective policy when the present U.S. administration is replaced after the next 2 to 6 years by its successor. The expanding legal regime in space will in any case benefit from the provision of maximum information to the American and world public and from maximum public discussion.

There is another, more hopeful, possibility; the protests and objections of foreign governments and military and commercial users of satellites may cause the United States to hold back from proceeding with the weaponization of space even while it resists pressures to negotiate a treaty prohibiting weaponization. This may be the best outcome we can hope for at this time, but it will require continued energetic opposition to weapons in space.

There is a possible comparison between the issue of weaponization of space and the emergence of nuclear weapons. Efforts to ban nuclear weapons completely gave way slowly to the efforts to tame these weapons. Efforts to control and tame space weapons are coming earlier in the cycle and space weaponization may emerge more slowly with a longer interval before the first use of these devices as weapons than was the time between Trinity and Hiroshima. Consequently, there may be more time to play out the recurrent contest between human capacity to invent new weapons and the efforts of human society to control them. Let us hope that this time is well used.

Chapter 15
Rethinking Security in the New Century: Return to the Grotean Pattern (2008)

15.1 Responding to 1989: Towards Cooperative Security

The main business of human society is to safeguard the life of its members.[1] This rich and fascinating volume surveys the many ways of protecting humankind against the threats to human life in today's world—armed conflict in all its forms, inhumane treatment, disease, natural catastrophe, the consequences of man-made environmental degradation, and scarcity of food, water and health care. The emphasis of the book is on the years since the end of the Cold War in 1989–1990, and on the challenges to security, old and new, with a special focus on environmental and human security, which have arisen in that period.

As we will describe further, a pattern of transatlantic cooperation among governments and civil society groups to cope with security challenges began to emerge in Europe after the Napoleonic Wars. With important exceptions, this pattern continued in the 19th and 20th centuries and into the post-cold war period. This trend confirmed Grotius' analysis of the human condition—the global nature of human society, its solidarity in agreeing on rules and new forms of cooperation to meet challenges to human life, and its emphasis on the importance of individuals and groups as well as of states, which—despite devolution of their powers

[1] This text was first published as: "Rethinking Security in the New Century—Return to the Grotean Pattern", in: Brauch, Hans Günter; Oswald Spring, Úrsula; Mesjasz, Czeslaw; Grin, John; Dunay, Pal; Behera, Navnita Chadha; Chourou, Béchir; Kameri-Mbote, Patricia, Liotta, P.H. (Eds.): *Globalization and Environmental Challenges: Reconceptualizing Security in the 21st Century*. Hexagon Series on Human and Environmental Security and Peace, vol. 3 (Berlin—Heidelberg—New York: Springer-Verlag, 2008): 3–6. Permission to republish this text was granted by Springer in Heidelberg and by AFES-PRESS on 8 March 2014.

H. G. Brauch and T. Grimwood (eds.), *Jonathan Dean*,
SpringerBriefs on Pioneers in Science and Practice 19,
DOI: 10.1007/978-3-319-06662-2_15,

to supra- and sub-national entities—remain the main units of the international system.

The events of the years since 1989–1990 have in general shown a worldwide trend of cooperation in dealing with man- and nature-made crises. They have largely repudiated the Hobbesian use-of-force approach. At the same time, they have provided renewed evidence that the world is not ready for a central governmental authority. Although efforts to control war showed some improvement in this period, attempts to deal with human-caused environmental degradation made little progress in blocking a process which in time may make this planet uninhabitable for human population. Rapid increase of that population is one cause of the problem.

15.2 New Security Challenges: Unilateral American Responses

The major events of the years since 1989–1990 included a worldwide cooperative effort in the 1991 Gulf War to repulse the Iraqi invasion of Kuwait. The widespread terrorist attacks on Western and other targets from the early 1990s to the present have failed to bring the popular uprisings in the Muslim states in support of the fundamentalist cause hoped for by terrorist leaders. However, they did elicit worldwide anti-terrorist cooperation of police, intelligence, and finance control, and the beginnings of cooperative efforts to deal with some of the underlying causes of terrorism. Fears of terrorist use of WMD remain widespread, although in fact the main terrorist weapon has remained conventional high explosives.

The U.S. military action in Afghanistan following the 9/11 attacks on New York and Washington, although deliberately unilateral and refusing many offers of help, was quite widely supported in world opinion. However, the U.S.–UK military action against Iraq in March 2003 broke radically with the pattern of cooperative engagement of previous U.S. administrations. It showed the costly effects of a go-it-alone policy, including inability to use the international institutions—UN weapons inspection and the Security Council—which might have neutralized the Iraqi regime. The consequence was U.S. inability to elicit more than token military, political, and economic cooperation in dealing with Iraq, capped by unwillingness of the Bush administration to devote the military and economic resources needed to cope with the task in Iraq. This outcome clearly showed the limits of U.S. 'super-power' and the unambiguous need for a cooperative approach.

15.3 Grotius on Preventive Attack

It is interesting to recall that wide international disapproval of the Bush administration's doctrine of preventive attack had been foreshadowed by Hugo Grotius (1625), when he said "to maintain that the bare probability of some remote or future annoyance from a neighboring state affords a just grounds of hostile aggression, is a doctrine repugnant to every principle of equity." (*On the Law of War and Peace*, Book II, Chapter I, para. XVII).[2] Pointing to the crucial difficulty of obtaining accurate intelligence about an adversary's intentions, Grotius points out that action in self-defense is not justified "unless we are certain, not only regarding the power of our neighbor, but also regarding his intention." (Book II, Chap. 22, para. IV).

15.4 Natural Disasters of 2004/2005 and Cooperative Security Responses

Natural catastrophes in the form of the December 2004 Tsunami in the Indian Ocean, equally devastating hurricanes in the Gulf of Mexico in late summer 2005, and a huge earthquake in Kashmir and Northern Pakistan in October of 2005 brought cooperative efforts to temper the disasters. There was during 2005 worthwhile cooperation between the U.S. government, WHO, the EU, and Asian governments in preparing defences against the avian flu. After long delays in each case, the United States joined Japan, South Korea, Russia and China in negotiating to curb the nuclear capabilities of North Korea, and with the UK, France and Germany in seeking to prevent development of nuclear weapons by Iran.

But the devastation of New Orleans and the Gulf Coast revealed the existence of an underprivileged underclass, while in November 2005, youth riots in Muslim suburbs of French cities suddenly exploded out of years of low social regard and extremely limited job and career opportunities and could portend serious confrontations ahead.

---

[2] See: Grotius, Hugo, 1975: *Prolegomena to the Law of War and Peace* Indianapolis: Bobbs-Merril); Grotius, Hugo, [2]1990 [1625]: "Prolegomena to the Law of War and Peace" in: Vasquez, J.A. (Ed.): *Classics of International Relations* (Englewood Cliffs, NJ: Prentice Hall); Grotius, Hugo; Tuck, Richard, 2005: *The Rights of War and Peace* (Natural Law and Enlightenment Classics) (Paperback); on Grotius: Bull, Hedley; Kingsbury, Benedict; Roberts, Adam (Eds.), 1992: *Hugo Grotius and International Relations* (Oxford: Clarendon Press); Edwards, Charles S., 1981: *Hugo Grotius. The Miracle of Holland. A Study of Political and Legal Thought* (Chicago: Nelson-Hall); Onuma, Yasuaki (Ed.), 1993, 2001: *A Normative Approach to War. Peace, War, and Justice in Hugo Grotius* (Oxford: Clarendon Press); Tuck, Richard, 2001, 2005: *The Rights of War and Peace: Political Thought and the International Order from Grotius to Kant* (Oxford: Oxford University Press).

15.5 Two Hundred Years of Cooperative Security

I have mentioned the emergence of cooperative efforts to control war in the Napoleonic period. Two hundred years ago, as the Napoleonic wars were bringing casualties of millions and huge political disruption, a new phenomenon emerged in the history of war. It consisted of two components. The first was establishment of multinational public peace societies proposing a wide range of institutions for avoiding or controlling war, like compulsory arbitration by a neutral international umpire and agreed limitation of arms.

Often in history there has been intense public opposition to specific wars, for example, the opposition in Russia to continuing World War I which led to the Bolshevik Revolution, and the opposition to the Vietnam War in the United States and elsewhere. But what happened in the early 19th century after acceptance of war over millennia as desirable or at least as a given component of human history, was the emergence of organizations which categorically opposed war as such. The names and dates of the new organizations in the U.S. and UK were significant: *The Massachusetts Peace Society* (1814), the *New York Peace Society* (1815), the *London Peace Society* (1816), and the *American Peace Society* (1828). These associations agitated for peace and against war through public meetings, pamphlets and tracts, and by lobbying with governments. From the outset, and throughout the 19th century, these associations collaborated with organizations in the United States on the one hand and organizations in Great Britain, France, Belgium and Germany on the other, forming a transatlantic community of peace interests. The Western European peace associations were from the outset sceptical of the efforts to achieve categorical rejection of war energetically pursued by the Americans, preferring to promote specific measures to avoid or limit war.

15.6 Cooperative Security Since the Vienna Final Act of 1815

The second component was the radical innovations of ongoing cooperation among the victors in war, in this case the victors over Napoleon, to maintain the peace. A large part of the credit for this change was due to far-sighted British policy. Prime Minister William Pitt the Younger began to plan the post-war peacekeeping structure in the 1790s. British cash was used to pay off the other main victors over Napoleon—the governments of Austria, Prussia, and Russia—to keep them engaged in the peace process. The four governments formed the Quadripartite Alliance and negotiated the 1815 Vienna Final Act setting forth the terms of the European peace settlement. The British urged that representatives of the four victorious powers meet periodically to discuss and decide on issues arising from the implementation of the Vienna Final Act and to ensure the peace of Europe. To keep a friendly eye on France and to engage French resources in the post-war settlement, France was later admitted to the Quadripartite Alliance, much as defeated Federal Germany was admitted to the NATO alliance over a century later. Continuing Pitt's far-sighted cooperative approach to security, British Foreign Secretary Canning extended to the Western hemisphere a prohibition against territorial acquisition by European states. Cooperation between the British Navy and a much weaker U.S. Navy created a transatlantic zone of peace.[3]

Over the years, European and American peace associations and governments collaborated in a series of agreements limiting war, like the 1856 Paris Declaration Respecting Maritime Law, the first Geneva Convention (1864) and the agreements at the first and second Hague Peace Conferences. The Concert of Europe lasted only until 1822 in its full form, but for many years peacetime coordination by ambassadors and senior officials continued and reached many agreements. The important innovation of ongoing peacetime coordination of international security by the victors in war was replicated and expanded by the victors in World War I and World War II in the form of the League of Nations and of the United Nations.

As we have seen, by the middle of the 19th century, Western—i.e. American, British and Western European—governments and public associations were nagged in close dialogue, exchange of ideas, and in intermittent collaboration on preventing and controlling war and were establishing institutions and treaties to this end. In fact, a rudimentary global security system was emerging through this transatlantic collaboration.

This collaboration continued throughout the 19th century. And, in fact, despite, or because of, the failure of World War I and of World War II, it continued through the 20th century.

This is not the place to attempt to describe the reasons why, after thousands of years of warfare throughout human history, a revolutionary change in public and also governmental attitudes toward war began to emerge in the early 19th century, but at least some of the underlying causes for this radical change seem evident. They include: (1) technological weapon innovation and the mounting carnage, destruction, and cost of war; (2) modern communications and media, which rapidly brought news of military events to publics as well as government officials; (3) social factors, including rising levels of income and education after the Industrial Revolution—this broadened the intellectual horizons of governmental officials and encouraged participation of publics in issues of war and peace; (4) changing, shared values of government officials and publics. These included the emergence of the Grotean idea of a known planet occupied by members of a single species. Finally, (5) the growth of democratic governments and institutions enhanced the influence of the electorate on security and other issues, and the openness of

---

3 See e.g.: Holsti, Kalevi J., 1991: *Peace and War: Armed Conflicts and International Order 1648–1989* (Cambridge: Cambridge University Press); Kissinger, Henry A., 1994: *Diplomacy* (New York: Simon & Schuster); Osiander, Alexander, 1994: *The States System of Europe. 1640–1990. Peacemaking and the Conditions of International Stability* (Oxford: Clarendon Press).

governments to public opinion. Growing understanding and cooperation in the especially difficult area of controlling war and armed conflict was accompanied by the growth of a habit of international cooperation in coping with natural disasters.

15.7 Shift from Cooperative to Unilateral Security Policy?

The trend toward global cooperation in a wide variety of areas was continued after the end of the Cold War by skilful diplomacy in the administration of George H.W. Bush, with the unification of Germany and the dissolution of the Soviet Union and the first Gulf War. But the trend toward increasing international cooperation was then sharply broken by the second Bush administration, intoxicated by its situation as the sole superpower, and determined to wield its power without the limitations imposed by allies.

To find the reasons for this sharp break in U.S. policy, we have to go back to the foundation of the United States in revolution against established power, its population by political refugees of all kinds, and to the growth of the concept that the United States were especially favored by divine providence in its institutions and values. At the outset of the 20th century, a large (25 %) component of the American electorate was characterized by attitudes of suspicion and superiority to the outside world and a desire to be isolated from it. But for over 75 years, from World War I to the end of the Cold War, public manifestation of the isolationist position in the U.S. was considered unpatriotic and suppressed through public disapproval. During this period, the reality of the outside world and of American military power became evident to all, including the isolationists. The end of the Cold War removed the pressure of public disapproval and abruptly released the pent-up forces of American isolationism in the transmuted form of heavily armed, highly nationalistic unilateralism, which captured control of the Congress in 1994 and of the presidency in 2000.

15.8 Returning to the Cooperative Tradition of Security Policy

Policy errors, military reverses, denial of cooperation by foreign governments, and the growing disaffection of the American electorate have tempered some of the hubristic excesses of the administration of George W. Bush. The chances are good that after one or two congressional election cycles and a presidential election, the United States will rejoin its own cooperative tradition of the past century and that the trend toward a cooperative world security system will be resumed, with greater U.S.–European collaboration at the UN, in controlling armed violence, and in coping with the environment.

Jonathan Dean with his wife Theodora in Mesa, Arizona in March 1999. *Source* Private photo collection of the Dean family

Amb. Jonathan Dean speaking with third graders at Coronado Elementary School in Gilbert, Arizona. March 3, 2012. *Source and copyright* Private photo collection of the Dean family

Ambassador Jonathan Dean: 1924–2014

Obituary, David Wright, physicist & co-director, Global Security, January 27, 2014

The arms control community lost one of its true heroes recently with the passing of Ambassador Jonathan Dean. My colleagues and I at UCS remember him especially fondly.

After retiring from a long career in the U.S. Foreign Service, Jonathan came to UCS at age 60—with no thought of slowing down—so he could continue to work on the issues that motivated him throughout his life. He spent more than 20 years here, bringing the same commitment to peace, and the tireless and patient approach to his work, that he showed during his government career.

Jonathan's interest in the security field was in no way theoretical. He was a wiz-kid—he started at Harvard at age 16—but dropped out the next year after the bombing of Pearl Harbor to join the army. That entailed going to Canada since joining the U.S. army at his age required permission from his parents, who were dead set against his plan. In June 1944, he landed at Normandy with Canadian forces 2 days after the initial U.S. D-Day landing. His chin carried a scar from a bullet wound he received later in the war. In the early 1960s, after joining the State Department, he served in the Congo at a time of turmoil at significant personal risk.

I was fortunate to have spent more than a decade working as a colleague of Jonathan's at UCS. But a few more personal recollections stick out in my mind.

First, Jonathan was one of the least cynical people I've ever known—despite his being a very acute observer of the real world. Over the years I've found that to be especially rare in Washington. He seemed to accept the frustrations and slow progress inherent in trying to advance arms control, including slogging through 8 years as the negotiator of the unsuccessful U.S.–Soviet talks to limit conventional forces in Europe (the Mutual and Balanced Force Reduction talks). But he saw this work eventually bear fruit when the follow-on talks—the Conventional Forces in Europe (CFE) Treaty—ended up cutting conventional forces far beyond what anyone imagined possible.

The second thing I remember is that he never seemed to be discouraged. He would craft policy proposals, patiently explain them, and carefully listen to the responses. Based on that he would re-craft his proposals and start the cycle again.

H. G. Brauch and T. Grimwood (eds.), *Jonathan Dean*,
SpringerBriefs on Pioneers in Science and Practice 19,
DOI: 10.1007/978-3-319-06662-2,

He knew that part of the battle was educating people and supplying them with motivation and good ideas, and that part of it was outlasting people or situations that were barriers to progress.

The third thing is that Jonathan would engage—respectfully—with anyone who was serious about grappling with the issues he cared about, regardless of where they were on the policy or political spectrum. He had friends in DC who were much more conservative than he was. I also remember on several occasions that he flew to Germany to address German peace groups that were considered too 'fringe' by most policy-types to be taken seriously. He was for years an advisor to and collaborator with Randy Forsberg, who started the Nuclear Freeze movement. Jonathan was one of the most decent and committed people I've met. He was a remarkable man and a true inspiration.

AG Friedensforschung und Europäische Sicherheitspolitik Peace Research and European Security Studies

A F E S P R E S S

Peace Research and European Security Studies (AFES-PRESS) was founded in September 1987 as an international scientific society under German law.

Aims of AFES-PRESS e.V.

- Scientific research projects: since 1987 on international relations, peace research, on security policy, since 1995 on environment, energy and climate policies and since 2001 on the global rethinking of security and since 2012 on sustainability transition and sustainable peace.
- Organization of scientific conferences and workshops
- Publication of the research results as AFES-PRESS Reports (non-profit publishers)
- support of international scientific co-operation and of young scientists

Board Members

- *PD Dr. habil. Hans Günter Brauch*, chairman (political science, international relations), Mosbach, Germany.
- *Prof. Dr. John Grin*, deputy chairman, University of Amsterdam (physicist, political scientist), Netherlands.
- *Assoc. Prof. Dr. Czeslaw Mesjacz*, deputy chairman, Economic University of Krakow (physicist, economist), Poland.
- *Prof. Dr. Ursula Oswald Spring*, chair of the Scientific Advisory Board, National Autonomous University of Mexico (UNAM)
- *Prof. Dr. Jürgen Scheffran*, Hamburg University, Germany.
- *Mr. Thomas Bast,* cashier, Germany.

Scientific Advisory Board

AFES-PRESS has an international Scientific Advisory Board that was chaired and co-chaired by *Ambassador Dr. Jonathan Dean* (1991–2014) and since 2007 by *Prof. Dr. Ursula Oswald Spring*,.

H. G. Brauch and T. Grimwood (eds.), *Jonathan Dean*,
SpringerBriefs on Pioneers in Science and Practice 19,
DOI: 10.1007/978-3-319-06662-2, © The Author(s) 2014

Research Projects and Conferences

Since 1987 AFES-PRESS has conducted research projects on: Armaments dynamics and East-West-conflict in the nuclear age; Institutions, procedures and instruments of a preventive arms control; several studies for the German Environment Office (UBA); Armament Policy of the United States since the End of the Cold War, and a multi-volume global publication project that has been published with Springer Publishers since 2003.

AFES-PRESS e.V. has organized several international scientific conferences on: Space Weapons (1985); SDI and Europe (1985); Aspects of European security (1986); Weapons technologies—arms control (1987); Military doctrine and arms control (1988); Verification and arms control (1989); Military research and development and disarmament (1991); Confidence building and partnership in the Western Mediterranean (1993) as well as major workshops at the Pan European Conferences on International Relations in Heidelberg (1992), Vienna (1998), Canterbury (2001), at ISA in Montreal (2004), at IPRA in Sopron (2004), in The Hague (2004) at WISC in Istanbul (2005), and at the Sixth Open meeting of IHDP (2005). AFES-PRESS maintains two websites: http://www.afes-press.de and http://www.afes-press-books.de

Contact Details

PD Dr. Hans Günter Brauch, Alte Bergsteige 47, D-74821 Mosbach, Germany
Phone: +49-6261-12912 Fax: +49-6261-15695 Email: brauch@afes-press.de and brauch@onlinehome.de

AFES-PRESS Obituary Ambassador Dr. Jonathan Dean (1924–2014)

Ambassador Dr. Jonathan Dean has peacefully passed away on 14 January 2014 in the age of 89. Ambassador Jonathan Dean was a distinguished American diplomat and a concerned U.S. citizen who after his retirement from the State Department in 1981 joined the Carnegie Endowment for International Peace as resident associate for arms control and European security issues, since 1984 he was global security adviser of the Union of Concerned Scientists (UCS) and he was associated with many other organizations in the United States (former president of the United Nations Association of the Washington; president, Institute for Defense and Disarmament Studies; Research Advisory Board, Project on Defense Alternatives; Advisory Board, Global Security Institute; Director, NGO Committee on Disarmament, Peace and Security; Steering Committee Member, Global Action to Prevent War; National Advisory Board, Council for a Livable World) and in Europe (Council Member, British American Security Information Council) working for détente, arms control and disarmament. Since 1991 he was chairman and since 2007 co-chair of the scientific advisory council or Peace Research and European Security Studies (AFES-PRESS), a Germany-based international non-profit scientific society.

He authored several books and his views were published in the *New York Times, Los Angeles Times, Boston Globe, Christian Science Monitor, Washington Post* and *Baltimore Sun,* as well as in journals such as *Foreign Policy, International Security* and *Foreign Affairs.*

As a diplomat Jonathan Dean was actively involved in integrating Germany into the West and gradually overcoming the Cold War, as the Deputy U.S. negotiator of the Berlin Agreement (1971) and as deputy and U.S. representative at the MBFR talks. After his retirement he contributed conceptually to the development of policies that aimed at détente in Europe, arms control and global disarmament prior and after the global turn.

Amb. Dean participated as a keynote speaker at several international conferences organized by AFES-PRESS in December 1989 and in October 1991 in Mosbach and he also contributed to several books and AFES-PRESS Reports. His last text will be published in Chinese in 2014 to commemorate his 90th birthday.

H. G. Brauch and T. Grimwood (eds.), *Jonathan Dean*,
SpringerBriefs on Pioneers in Science and Practice 19,
DOI: 10.1007/978-3-319-06662-2,

"Rethinking Security in the New Century—Return to the Grotean Pattern", will be published in 2014 in Chinese in: Hans Günter Brauch, Úrsula Oswald Spring, Czeslaw Mesjasz, John Grin, Liu Cheng (Eds.): *Globalization and Environmental Challenges: Reconceptualizing Security in the 21st Century* (Nanjing: Nanjing Press Company, 2014).

The members of the board of AFES-PRESS will remember the many conceptual discussions with Amb. Dean and his intellectual contributions that were ahead of official policies. He will be remembered as a nice colleague and as a fatherly friend.

Board of Peace Research and European Security Studies (AFES-PRESS) e.V.

PD Dr. phil. habil. Hans Günter Brauch, Germany, Chairman

Prof. Dr. John Grin. University of Amsterdam, The Netherlands, Deputy chairman

Prof. Dr. Czeslaw Mesjasz, Economic University of Cracow, Poland, Deputy chairman

Prof. Dr. Ursula Oswald Spring, National Autonomous University of Mexico member of the AFES-PRESS board, and chair woman of its Scientific Advisory Commitee

Prof. Dr. Jürgen Scheffran, Hamburg University, Germany

Thomas Bast, Cashier, Germany

Mosbach/Bangkok, Amsterdam, Cracow, Cuernavaca, Hamburg, Owen, 20 January 2014

About the Editors

Hans Günter Brauch (Germany), Dr., Adj. Prof. (Privatdozent) at the Faculty of Political and Social Sciences, Free University of Berlin (ret.); since 1987 chairman of *Peace Research and European Security Studies* (AFES-PRESS). He is editor of the *Hexagon Book Series on Human and Environmental Security and Peace* (HESP), and of *SpringerBriefs in Environment, Security, Development and Peace* (ESDP) and of the *SpringerBriefs on Pioneers in Science and practice* with Springer-Verlag, He was guest professor of international relations at the universities of Frankfurt on Main, Leipzig, Greifswald, and Erfurt; research associate at Heidelberg and Stuttgart universities, and a research fellow at Harvard and Stanford Universities.

In fall and winter 2013/2014 he was a guest professor at Chulanlongkorn University in Bangkok (Thailand). He published on security, armament, climate, energy, and migration, and on Mediterranean issues in English and German, was translated into Spanish, Chinese, FrenchRussian, Japanese, Danish, Finnish, Greek, Portuguese, Serbo-Croatian, Turkish. Recent books in English: (co-ed. with Liotta, Marquina, Rogers, Selim): *Security and Environment in the Mediterranean. Conceptualising Security and Environmental Conflicts*, 2003; (co-ed. with Oswald Spring, Mesjasz, Grin, Dunay, Chadha Behera, Chourou, Kameri-Mbote, Liotta): *Globalization and Environmental Challenges: Reconceptualizing Security in the 21st Century, 2008;* (co-ed. with Oswald Spring, Grin, Mesjasz, Kameri-Mbote, Chadha Behera, Chourou, Krummenacher): *Facing Global Environmental Change: Environmental, Human, Energy, Food, Health and Water Security Concepts* (2009); (co-ed. with Oswald Spring): *Reconceptualizar la Seguridad en el Siglo XXI* (2009); (co-ed. with Oswald Spring, Mesjasz, Grin, *Kameri-Mbote*, Chourou, Dunay, Birkmann): *Coping with Global Environmental Change, Disasters and Security—Threats, Challenges, Vulnerabilities and Risks* (2011); (co-ed. with Scheffran, Brzoska, Link, Schilling): *Climate Change, Human*

H. G. Brauch and T. Grimwood (eds.), *Jonathan Dean*,
SpringerBriefs on Pioneers in Science and Practice 19,
DOI: 10.1007/978-3-319-06662-2,

Security and Violent Conflict—Challenges for Societal Stability; (co-ed. with Oswald Spring, Tidball): *Expanding Peace Ecology: Peace, Security, Sustainability, Equity and Gender: Perspectives of IPRA's Ecology and Peace Commission* (2014); (co-ed. with Oswald Spring, Grin, Scheffran): *Handbook on Sustainability Transition and Sustainable Peace* (2015).

Address: PD Dr. Hans Günter Brauch, Alte Bergsteige 47, 74821 Mosbach, Germany.
Website: http://www.afes-press.de and http://www.afes-press-books.de/.
Email: brauch@afes-press.de.

Teri Grimwood (USA) is a researcher with the Global Security Program of the Union of Concerned Scientists. In 1993 she came to UCS to provide research and editorial support to Ambassador Jonathan Dean, and worked closely with him for the next 12 years. She maintained contact with both Ambassador Dean and his family after his departure from UCS.
Address: Teri Grimwood, Union of Concerned Scientists of the USA, *Washington, DC, Office*, 1825 K St. NW, Ste. 800, Washington, DC 20006-1232, USA.
Website: http://www.ucsusa.org/about/staff/staff/teri-grimwood.html.
Email: TGrimwood@ucsusa.org.

About the Author

Jonathan Dean (United States of America): a former ambassador and since 1984 Adviser on Global Security Issues, Union of Concerned Scientists (1984–2007). He has a Ph.D. in political science from George Washington University. During World War II, he saw combat infantry service from Normandy to the Elbe. He joined the U.S. Foreign Service in 1949, where he worked mainly on issues of East-West relations, disarmament, and international peacekeeping. In 1950 in Bonn he was liaison officer between the U.S. High Commission and the Federal German government, served as desk officer for East Germany, and then as first secretary at the U.S. Embassy in Prague.

In 1968 he returned to the U.S. embassy in Bonn as deputy U.S. negotiator for the 1971 quadripartite agreement on Berlin, and from 1978–1981 served as U.S. negotiator in Vienna on the Mutual and Balanced Force Reductions (MBFR) talks between NATO and the Warsaw Pact. He authored: *Watershed in Europe* (1987), *Meeting Gorbachev's Challenge* (1989), *Ending Europe's Wars* (1994), and was a co-author of *The Nuclear Turning Point* (Brookings, 1999). His views have been published in major American newspapers and foreign policy journals.

Address: Union of Concerned Scientists, 1825 K Street NW, Suite 800, Washington, DC 20006, USA.

Email care of: tgrimwood@ucsusa.org and for the Dean family tdschiro@gmail.com. See also the website on this book, at: http://www.afes-press-books.de/html/SpringerBriefs_PSP_Dean.htm.

H. G. Brauch and T. Grimwood (eds.), *Jonathan Dean*,
SpringerBriefs on Pioneers in Science and Practice 19,
DOI: 10.1007/978-3-319-06662-2,

About the Book

Jonathan Dean (1924–2014) was a distinguished U.S. diplomat (1950–1980) and from 1984 to 2007 global security adviser to the Union of Concerned Scientists, Washington, D.C. During the 1980s and 1990s he was a pioneering conceptual thinker, writer and speaker on détente in Europe, global cooperative security, arms control and disarmament. He authored: *Watershed in Europe* (1987), *Meeting Gorbachev's Challenge* (1989), and *Ending Europe's Wars* (1994). This volume contains his biography and bibliography, six texts on detente and arms control in Europe in the 1980s: Beyond First Use; MBFR; Alternative Defence; Berlin; A Crisis Management Center; Conventional Arms Control in Europe; and six texts on the new security order in Europe since the 1990s: Components of a Post-Cold War Security System for Europe; Constraining Technological Weapons Innovation; NATO Enlargement: Act II; Ten Years after the Wall; Future Security in Space; Rethinking Security: Return to the Grotean Pattern.

H. G. Brauch and T. Grimwood (eds.), *Jonathan Dean*,
SpringerBriefs on Pioneers in Science and Practice 19,
DOI: 10.1007/978-3-319-06662-2,

MIX
Papier aus verantwortungsvollen Quellen
Paper from responsible sources
FSC® C105338

If you have any concerns about our products,
you can contact us on
ProductSafety@springernature.com

In case Publisher is established outside the EU,
the EU authorized representative is:
Springer Nature Customer Service Center GmbH
Europaplatz 3, 69115 Heidelberg, Germany

Printed by Libri Plureos GmbH
in Hamburg, Germany